*f*P

DIRTY MINDS

How Our Brains Influence Love, Sex, and Relationships

Kayt Sukel

Free Press
New York   London   Toronto   Sydney   New Delhi

*f*P

FREE PRESS

A Division of Simon & Schuster, Inc.

1230 Avenue of the Americas

New York, NY 10020

First Free Press hardcover edition January 2012

FREE PRESS and colophon are trademarks of Simon & Schuster, Inc.

For information about special discounts for bulk purchases, please contact Simon & Schuster Special Sales at 1-866-506-1949 or business@simonandschuster.com.

The Simon & Schuster Speakers Bureau can bring authors to your live event. For more information or to book an event contact the Simon & Schuster Speakers Bureau at 1-866-248-3049 or visit our website at www.simonspeakers.com.

Designed by Julie Schroeder

Manufactured in the United States of America

1 3 5 7 9 10 8 6 4 2

Library of Congress Cataloging-in-Publication Data

Sukel, Kayt.
Dirty minds : how our brains influence love, sex, and relationships / Kayt Sukel.
p. cm.
1. Love—Psychological aspects. 2. Love—Physiological aspects.
3. Sex (Psychology) 4. Sex (Biology) I. Title.
BF575.L8S9125   2012
155.3'1—dc23
2011025606

ISBN 978-1-4516-1155-7
ISBN 978-1-4516-1158-8 (ebook)

To Chet, my mother, and all those cheeseburgers

Contents

—

A Note about Illustrations

—

The brain is an organ with an intricate biological architecture. When discussing the results of neuroimaging studies, I have used illustrations to help point out the approximate location of key areas in the brain. However, due to the brain's complexity, it can be difficult to differentiate all the areas of interest in a single study from the same visual perspective. In some cases, only the most significant regions are highlighted, or two images are used. In addition, laterality (whether the activation occurred on the left hemisphere, the right hemisphere, or both hemispheres of the brain) is often ignored for simplicity's sake. For those interested in the areas not illustrated or in more exact positioning, I recommend visiting the Whole Brain Atlas, a detailed online neuroimaging primer created by Keith A. Johnson, MD, and J. Alex Becker, PhD, www.med.harvard .edu/AANLIB/home.html.

What is this thing called love?

—COLE PORTER

Introduction

—

We all know what love is. Or at least we think we do. Love is a rock, a drug, reciprocal torture, and an exploding cigar. Love is all you need—yet it's a cold and a broken Hallelujah. It is a many-splendored thing, a battlefield, and a river. Love stinks. It's divine. It is never having to say you're sorry. Or perhaps having to say so much more than you ever thought you would. Love is a bitch. It's like a disease. It's a trap. Ted Nugent, in one of his more romantic moments, even likened love to a tire iron. There are plenty of metaphors out there, and many ring true, but there is still no great, all-encompassing definition. Perhaps that is why someone thought it necessary to create a bumper sticker that simply says, "Love is . . ." and avoid the specifics altogether. Try it yourself some time—explain the concept of love. Surely *something* comes to mind. Now make it applicable to *everyone* in *every* potential love-related situation. It isn't easy, is it?

"It's like what the Supreme Court said about pornography—I know love when I see it!" a friend of mine suggested. "Or feel it, as it were."

He definitely had something there. I think most of us innately *know* what love is. We can recognize it and feel it. We just cannot translate it into words. Love is simply too abstract, too elusive, and too bizarre to explain. The same is true of love's partners in crime: sexual attraction, lust, monogamy, and hate. Anything as complicated as love is outside the realm of simple description—and best left to philosophers, novelists, and boy bands.

The lack of a clear-cut definition has not stopped folks from offering advice on how to attract, nurture, and prolong love. Moms, friends, even total strangers are happy to tell you all about the *right* ways to handle your love life. They are usually tips bearing the promise of relationship rescue through better understanding, better communication, and better sex. Never mind that far too much of this guidance falls into the "Take my advice, I'm obviously not using it" category. Anecdotal evidence of the communication strategies (or crazy bedroom antics) that fixed Aunt Deirdre and Uncle Mike's marriage just do not cut it anymore. In this day and age we want our advice—even for something as intangible as love—backed by cold, hard science. Instead of Mom's shoulder, Freud's couch, or the pastor's office, we now look for answers in genetic profiles and brain scanners.

I once saw a television commercial for an acne drug. The tagline was "Blame biology." Forget diet, proper hygiene, or a good bar of soap (my own dermatologist's go-to solution): this commercial insinuated that acne was solely a biological issue and that perfect skin was only a doctor's prescription away. Advances in research mean that previously inexplicable phenomena like depression, obesity, and a whole host of other genetic disorders can now be examined within the realm of biology and are mostly treatable by this or that new pill. These pharmacological treatments are a boon for those of us who feel we are working too hard (and without much effect) to maintain some semblance of balance in our body and mind. We are made to believe that these sorts of problems are not our fault, that the blame falls squarely on that blasted biology. So biology should give us a way to fix it.

Intuitively it feels as though we should view love through the same sort of lens. Though some might argue the point, I consider myself a relatively intelligent person. Despite several long-term relationships (and more than a few short-term ones), a failed marriage, and a child, as I get older I realize that I do not know much about love. Some days I'm fairly certain I know absolutely nothing about it. And I do not seem to be alone in this. Start a conversation with someone falling in or out of love and nine times out of ten you will eventually hear the words "I should know better."

Most of us, if we're being honest, will admit we are a tad clueless when it comes to love, no matter how experienced we think we are. With that backdrop in place, blaming biology doesn't seem like such a bad option. Certainly we would not keep making the same silly mistakes if biology weren't urging us on.

Technology and science have now advanced to the point that disciplines like biology, genetics, epidemiology, evolutionary science, psychology, philosophy, computer science, and medicine have converged into the catchall field of neuroscience. More and more, neuroscientists are demonstrating that the brain *is* behavior—the two simply cannot be teased apart. Our brains are the seat of the biology that is directing every move we make. (I realize there are many who believe there is some kind of spiritual hand guiding our love lives, perhaps even influencing our brains. Given that it is a controversial argument that science can neither prove nor refute, I consider it outside the scope of these pages. But I will discuss some of the neuroscientific studies examining religious devotion and the brain in chapter 16.)

The latest neuroscientific discoveries provide a better understanding of the brain and its role in disease and behavior, including complex behaviors like attachment, romantic love, and sexual decision making. Researchers have now identified specific brain areas involved with love, different neurochemicals that may make us confuse love and lust, and genetic and environmental factors that may interact to change the way we approach our relationships. With these findings, the old-school debates about love framed in nature-versus-nurture scaffolding (that we can simply *decide* to change the way we act, as many self-help and dating manuals claim) are being countered by a new perspective suggesting that we humans may be more enslaved by our biology than was previously thought.

Of course, it is not quite as simple as that. Current research also points to complex ways in which our DNA is directly influenced by our environment. Epigenetics, a blossoming new field in neurobiology, is demonstrating that environmental influences can actually alter gene expression during development and beyond. As technology advances to allow a more focused examination of this intricate dance between our brains and our environment, we can pose new questions about the nature of love—questions drawn solely from neuroscience, the science of modern human biology—and shed old assumptions based on self-help, sociology, and spirituality.

With innovative, cutting-edge methods like neuroimaging, genome-wide association studies, and transgenic animal models, scientists now have the ability to observe love-related phenomena at the molecular level. Forget person-to-person communication and its importance in relationships; we can now measure the communication between brain cells.

"We're starting on a whole new world of research here," says Helen Fisher, an evolutionary anthropologist at Rutgers University who studies love from a scientific perspective. She has even shared some of her insights on the brain and love as a consultant to the online dating company Chemistry.com. "It is just beginning—the brain scanning, the epigenetic studies, tracing the molecular signaling pathways involved with love. There's so much here to help us answer the questions we've asked for centuries. We've only just begun to find out how love can turn off decision-making areas, how childhood may affect our emotional control, and for what kind of people love can become an addiction. Really, we're just beginning to truly understand the nature of love from a neurobiological perspective. It's tremendous."

This is certainly a novel way to look at a universal, enigmatic phenomenon. Many of us hope that these studies can provide the answers we have always sought concerning love: What is it, exactly? How can I make it last? Is monogamy natural, or even possible? What is it about that person—that person who is so *not* right for me—that I find so utterly irresistible? Why has my love for my child changed the way I care about everything and everyone else? There are many perplexing issues to choose from, and because love also has the power to make most of us feel like utter fools, a few good answers would be welcome. But can the examination of hormones, neural pathways, and epigenetic regulation of genes really give us the answers to the desperate queries of our insecure hearts?

To date there have been thousands of scientific inquiries about the nature of love. There are probably just as many love-related advice books that have had their tenure on self-help shelves in bookstores across the globe. These days many of those books (as well as the magazine articles and talk show themes based on them) claim to draw their advice from the newest neuroscientific evidence. They tell us that men and women have different kinds of brains and that we're hardwired for specific love-related behaviors. They say certain chemicals (some available on the Internet for only $19.95 plus shipping and handling) can help us attract the right mate. They suggest that our biology dictates *everything* and that heredity plays a huge role in whether we will find success in love. What's more, if we are willing to follow some simple guidelines based on the hottest brain research, we too can achieve relationship nirvana. The fact that the research findings cited in these books are often generalized, misinterpreted, or taken a wee bit out of context in order to promote a particular point of view does nothing to dampen their fans' enthusiasm. It's not such a surprise, really. These books tell us that if we follow the rules and approach our relationships in the "right" way, true love is within our reach. Some of these findings support ideas you may have had about love all along. Others seem wrong, *dead wrong*, in terms of both accuracy and appeal. They go against everything you were raised to believe about relationships. You read about those studies and find they have blown your mind (and heart) right out of the water, because if they are right, you are even more screwed than you thought: that true love thing is never going to happen for you. Headlines, guidelines, rules, lists, directions, and sound bites abound, but what can they really tell us about the nature of our hearts?

In fact neuroscientists offer no surefire way to avoid a broken heart or to make your marriage last. More often than not, they are far more interested in questions involving cognitive perception, consciousness, reward processing, or the behavior of previously uncharacterized genes. Some are looking at treatments for disorders like autism or cancer. The love stuff is important, yet somewhat tangential to the kinds of results scientists are really after.

Given the complex interplay between genes and environment in these behaviors (as well as the fact that most of these interactions are tested on animals and not people), any specific advice probably wouldn't do you much good even if it were offered. Neuroscience, particularly epigenetic inquiries into the brain, suggests that although people's brains have much in common, they may be just different enough to require individual treatments for their myriad diseases. By extension, there may be some individual differences in the ways we approach love, lust, and relationships.

"Behavior is a really complicated thing," says Alexander Ophir, a researcher at Oklahoma State University who studies mating and sexual behaviors in prairie voles, a breed of small rodent that forms the foundation of most monogamy research. "There are major differences between us and the voles. We humans have consciousness and culture that affect our behaviors. Taken all together, those things make the study of these behaviors quite messy."

More and more it is looking as if there is no one-size-fits-all approach to successful relationships. If we can, indeed, "blame biology," we can look only as far as our own. This idea is terrifying but liberating. Our complicated behaviors make for complicated minds—I'd even go so far as to say "dirty" minds, with so many variables muddying the proverbial waters—and make for complex study.

The following pages will provide no advice or guidelines regarding love. This book will not tell you how to become more attractive to the opposite (or same) sex, how to become a better parent, or even how to make your mate stay with you. And though I may want to tell you to drop the hot new relationship book, throw out those brain chemistry supplements you bought, or change the channel when Dr. So-and-So's syndicated advice show airs, I'll refrain.

What I will do is try to explain what neuroscience has actually learned about the various ways our brains can affect our hearts—and what those findings mean within the context of human behavior. Even without the inclusion of five steps to staying faithful or ten ways Mom can benefit from Dad's biological parenting style, it may put what you are reading or watching on television into the right context. I hope it will steer you away from picking up a pill or spray (or, if not, at least give you a better idea of what such a concoction might actually do). If nothing else, I hope to offer you a better understanding of why we humans act so strangely when it comes to that crazy little thing called love.

Chapter 1

The Neuroscience of Love: A History (Theirs and Mine)

In 1994 a scientist named Sue Carter submitted a grant application to study a hormone called oxytocin (not to be confused with the narcotic Oxycontin, aka hillbilly heroin) in a small rodent called the prairie vole.

A prairie vole family. *Photo by Todd Ahern, University of Massachusetts.*

A prairie vole (*Microtus ochrogaster*) looks a lot like your garden-variety mouse, but scruffier and with a shorter tail. Happily burrowing under gardens and meadows in a large stretch of central North America, these small rodents might completely escape our notice except for one special trait: they are monogamous.

Socially monogamous, that is. Unlike most other rodents—or most other mammals, for that matter—prairie voles form lifelong pair-bonds, or lasting social and sexual relationships with a single member of the opposite sex. Both males and females are also directly involved with the parenting of offspring. Because of the rarity of such habits in the animal kingdom, many animal behaviorists have become exceptionally interested in the prairie vole. One such researcher was Carter.

A professor of psychiatry at the University of Illinois at Chicago, Carter hypothesized that oxytocin, which is linked to childbirth and breastfeeding, could increase social attachment. She had already conducted research to support the idea and hoped that this grant would allow her to continue studying the hormone and its relationship to social behaviors in the prairie vole. In her application she did not mention love, marriage, or even humans. Somehow the grant review committee decided she was studying the little four-letter-word that begins with an *l*—love, that is—which was considered a serious no-no in the hard science climate of the day.

"I was trying to get federal grant funding to continue my work, and suddenly I was accused of studying *love*," she said when I visited her lab in Chicago. Petite, white-haired, and a little bohemian in style, Carter somehow managed the feat of being both incredibly welcoming and intellectually intimidating at the same time. "Honestly, it was a shock to me. I would not have used the word *love*—I never used the word *love*. I didn't think about the work in terms of love. I was simply talking about a preference of one animal for another—not some human construct that seemed to have little to do with what we were actually studying."

Carter told me she was unsure of how to respond to the review. She conferred with Kerstin Uvnäs-Moberg, a fellow scientist also interested in oxytocin who was working at Stockholm's Karolinska Institute. Could it be that their work was related to something as messy and indefinable as love? Might there be a neurobiological basis for the future study of love? Looking at newly published research by various labs concerning oxytocin, social attachment, and pair-bonding in prairie voles and other mammal species, the answer seemed to be yes. Carter and Uvnäs-Moberg thought it was time to stop ducking the topic and admit that their work did have implications for human behavior.

"It seemed like the time to really try to articulate and explain the idea that social bonds were critical to human love," Carter said. While sex was, is, and will be of the utmost importance to propagating our species, Carter and Uvnäs-Moberg were convinced that love needed to be articulated in the context not only of genetic propagation but also of survival—specifically, the ways social bonds can help people thrive in the face of stress and other complexities of life on a daily basis. Perhaps our brains promote social relationships in order to ensure that more than one person is on tap to avoid dangers, to make sure there is enough food around to feed the family, and to help raise the young'uns. Investigating how the neuroscience underlying social bonds might promote these behaviors seemed a pertinent line of study.

Though this line of inquiry seemed very clear to Carter and Uvnäs-Moberg, it was difficult to get respect (and, perhaps more important, funding) to study such ideas experimentally. There was already ample evidence in neuroscience literature to suggest that love was a worthy topic of research. But the scientists never called it such, avoiding it like the dirty word it is. Instead they referred to the related topics of pair-bonding, monogamy, attachment, and mating behaviors. If you read between the lines, there was a lot of information out there, perhaps even enough to make the neuroscientific study of love its own field. Still most professional scientists were afraid to call love by its true name.

There was no sense in talking about the neuroscience of love without a proper working definition—a common standard that scientists across disciplines could use to test and validate hypotheses. Sadly, as fitting (and poignant) as Ted Nugent's "tire iron" characterization might be as a song lyric, it would be limiting to use as the basis for a credible, replicable scientific study. To that end Carter and Uvnäs-Moberg invited thirty-eight prominent scientists in the field of neurobiology to a meeting at the 1996 Wenner-Gren Symposium in Stockholm titled "Is There a Neurobiology of Love?"

One of the products of that meeting was a definition. Instead of going with Merriam-Webster's basic statement about love being a case of "strong affection for another," the group consensus was that love is "a life-long learning process that starts with the relationship of the infant to his or her mother and the gradual withdrawal from the mother with a search for emotional comfort and fulfillment." This definition was included in the summary report written by the prominent neuroscientist Bruce McEwen.[1] It offers more detail than the definition of love as strictly an emotion or a basic mammalian drive, like hunger or thirst—even if it is less romantic than "sweet surrender" or "my first, my last, and my everything." Though a mouthful, this definition would serve as the standard to which future studies across the neurobiology field could refer.

The meeting also started a renaissance of sorts, a green light for neuroscientists, neurobiologists, and neuroendocrinologists to finally call love, well, *love*. This allowed them to start studying the nuances of this human phenomenon from the perspectives of brain and biology. Two years later many of the meeting's prominent attendees published studies in a special issue of the journal *Psychoneuroendocrinology* on topics ranging from the evolutionary antecedents of love to the physiological consequences of withholding it. With such respected scientists backing the concept, researchers could more easily study the *l*-word within the space of the brain and neurobiology.

Sexy Baby Banning

Fast-forward ten years. Many great studies concerning neuroscientific aspects of that "life-long learning process" we call love were published in the late 1990s, with a great number appearing in high-profile journals like *Nature* and *Science*. Brains, it seemed, have quite a bit to do with love—certainly far more than do our proverbial hearts. While working on a story for a neuroscience website, I accidentally stumbled across McEwen's meeting report. A simple misclick on a library database brought me to it, and even though it was completely off-topic I was compelled to read it.

Maybe I was drawn to the question raised in the title, "Is there a neurobiology of love?" It was not a subject I'd had occasion to study before. Maybe it was the fact that it was written by McEwen, an acclaimed neuroscientist from Rockefeller University. His work had impressed me since I was a graduate student. Maybe I was just procrastinating. I might well have read the phone book as an excuse to take a break that muggy afternoon. Or maybe it had something to do with my sleep deprivation. Did I mention I had recently become a mother?

If there is a stereotype of a new mom—think bedraggled, belea-guered, and baggy-eyed—I fulfilled it, and then some. From the stains on my shirt to the state of my house, there was not one part of my life left untouched by the effects of motherhood. As much as I do not subscribe to the notion of "mommy brain," or the idea that motherhood makes you stupid, I have to admit that I sometimes wondered what was going on upstairs. But honestly, what had changed the most—somewhat inex-plicably—since becoming a mother was my marriage.

The arrival of my son had completely altered my relationship with my husband. Though I certainly expected my marriage to change once we had children, I was not prepared for a complete loss of intimacy. We had been a tight-knit team, albeit a motley one, but now we were satel-lites in separate orbits, crossing paths only when it came to our child. My friends with kids assured me that the situation was natural and would right itself over time, after the shock of our new addition wore off. One friend, a mom of three, went beyond that: "You can't expect to feel the same way about your husband now. Your relationship needs to change so your son can be your focus. Our brains are wired so our kids can come first. It's an evolutionary thing."

Her statement stuck with me. I could not understand how an "evo-lutionary thing," as she had so eloquently put it, would rule out a nur-turing, loving relationship between two adults or an active sex life. Now that I had checked into the breeder category, wasn't I supposed to keep popping out kids to guarantee propagation of the ancestral line? Sex, if not a little passionate love, was required to fulfill that goal. Perhaps I had missed something.

It was a conundrum. Like most new moms, I was bone-tired. Yet I was enthralled by this small baby boy who somehow managed to brighten each moment of my life as he sucked the energy out of me. Like my friend's puzzling evolutionary edict, it was a contradiction I could not quite figure out.

As someone who wrote about neuroscience for a living, I began to wonder what role the brain played in what was happening to me. Maybe all of it—my crazy love for my son and my imploding marriage—could be explained by changes in my brain that occurred during and after childbirth. I remained convinced that my husband's brain may have been altered a little too. Picking up a copy of a meeting report about the neurobiology of love seemed as good a place as any to start learning more.

Reading McEwen's report, I was immediately struck by a single line, a quote McEwen added from British researcher and fellow Symposium attendee Nicolas Read: *"If we realized how sexy babies are they would have been banned."*

Frankly, that kind of sentence is an anomaly in the scientific literature. You do not find lines like that in many research papers. Trust me on this one. Most of the papers I read when I'm working on a neuroscience story include lines like, "Aβ deposition stimulates a local immune response by the microglia, which become macrophagic."[2] Though fascinating (once translated into plain human English, that is), it's not exactly the kind of stuff that makes you laugh out loud.

"It was a quip, obviously," McEwen said when I asked him why he decided to include it. "But the mother-child bond seems to be such a strong, remarkable phenomenon." What might researchers learn if they studied such things from a biological and mechanistic perspective? That is exactly what the attendees of the symposium wanted to start doing—as well as take a more mechanistic and biological approach to the study of monogamy, sex, and other love-related behaviors.

Even at such an early juncture in the neuroscientific study of love, McEwen had me at sexy baby banning. It has a certain perverse poetry to it. Not to mention that it resonates strongly for a new, sleep-deprived, and biologically curious mom like myself. To nonparents out there (and admittedly some of the parents too), I am sure this line sounds pretty creepy. Context is everything, after all. But to my mind, it fit. My baby was pretty sexy—much more so than I'd been prepared for. Not in a sweaty, naked-hot-guy kind of way, but in an irresistible, compelling way that had altered my body, my mind, and my life from top to bottom. I had no idea if these changes were due to evolution, neurobiology, or my particular situation, but I wanted to know more about how motherhood—and, yes, love—was facilitating them. So while I became more infatuated with my son each day, I was also just as intrigued by neuroscientific studies that gave me more insight into motherhood, monogamy, sex, and love.

Learning the In's and Out's of Love

I have already confessed that I know nothing about love. I've learned that you become much less reluctant to say so as soon as you are famil-iar with the inside of a divorce lawyer's office. The disintegration of my marriage was not sudden; it took years and years to circle the drain. You'd think there would have been ample opportunity to correct course during those few years when our unhappiness became apparent. But no matter how deeply I wanted to fix things, I just never found a chance. I doubt my former husband did either. Even in hindsight I cannot tell you where my husband and I went wrong. Just when I knew it would never again be right.

Apparently I have not mastered that "life-long learning process"; despite the fact that my baby, now a charming and curious kindergar-tener, remains as "sexy" as ever, my search for emotional comfort and fulfillment with a partner continues. Might it have something to do with my hormones? The way my brain is wired? My choice of partner? The length of time we were together? How often we had sex? The way my body, including my brain, changed after having a child? All of the above? It was something I needed to figure out in order to truly move on after my marriage fell apart, not to mention bolster me as I reentered the dat-ing realm.

Like most, I hoped to find some easy answers, some actionable advice that might help me understand my past relationships and, more important, avoid making the same mistakes in the future. With the dat-ing pool beckoning, I hoped that acquiring the right knowledge would make up for developing crow's feet, a postchildbirth figure, and postdi-vorce gun-shyness.

My quest to better understand the scientific nature of love started when I stumbled on a research paper and continued as I read all I could find on love, sex, and the brain. It was not enough to satiate my curiosity—I needed to talk to these scientists about their work and the different ways their findings might be interpreted. I had to visit a few labs and see some of the science in action. And surely it couldn't hurt to participate in a few studies myself.

I learned a lot over the course of this journey. I discovered that the study of love—in any manner, neuroscientific or otherwise—is a tricky, complicated thing, more so than a natural skeptic like me had ever realized. But despite many enlightening discussions and personal adventures, nothing I experienced offered any surefire intelligence on what the majority of us are hoping to figure out: how to find love and then keep it around for a while. As it so happens, there is no magic formula, no rule book for understanding the brain's role in love. There are no hard and fast answers. However, there are plenty of interesting surprises to be discovered about our "dirty" minds, both in animal models and in the fMRI scanner. To start I need to provide a little background, a taste of the brain areas and chemicals that fuel this most intoxicating of human emotions. This tale of brains in love begins with a bit of scientific history and the region of the brain responsible for reward processing called the basal ganglia.

———

Chapter 2

The Ever-Loving Brain

Could anything represent love better than a big red heart? Probably not. Even outside Valentine's Day (or rather, the six weeks of frenzied mass marketing that precede it) this particular image is omnipresent. So much so that replacing it with a brain, even now that we know the heart has little to do with emotion, seems just plain wrong. Can you imagine someone giving you a box of candy or an anniversary card shaped like a brain? It doesn't quite work. The idea is a bit disturbing even to a brain nerd like me.

Why is that one little symbol, a cardiovascular organ that we have now learned is basically just a blood pump, so pervasive? Perhaps it is because for the better part of a thousand years many of the best thinkers believed the heart was not only the seat of emotion, but also the center of rational thought and cognition.

We humans have wanted to differentiate ourselves as logical, reasonable beings since the earliest years of our history and have sought to denote a part of the body responsible for controlling this rationality. Aristotle believed that the heart was the source of human intelligence and vitality due to its heat and movement. (The brain, he believed, was responsible for cooling the passions created by said heart.) Plato, a teacher of Aristotle, disagreed with his student about the center of thought and emotion. He postulated the brain as the rational organ because it was the part inside the body closest to the heavens.

For years philosophers, theologians, and physicians volleyed back and forth about whether the heart or the brain was the most important organ (with the liver, of all things, occasionally making an appearance in the debate). But without access to scientific methods and the technologies with which to observe how these organs worked within the living body, there could be no definitive answer.[1]

Claudius Galen of Pergamum, a second-century Greek physician and philosopher, was one of the first to give more credit to the brain based on actual biological observations. This was likely due to his job: before he became the most influential physician of his time, he worked as a surgeon to the gladiators, where he surely witnessed ample evidence of what a good blow to the head could do to personality, movement, and behavior. In his treatise *De usa partium corporis humani (On the Usefulness of the Parts of the Body),* he argued that the *encephalon* (that's Ancient Greek for brain) must be responsible for both movement and perception. Otherwise it would not be attached to the sources of the senses (eyes, ears, nose, and mouth) as well as to the major motor nerves. He did not think the brain had anything to do with intelligence, per se—after all, plenty of stupid animals had fairly developed brains— but it was the organ that helped us process sensory input and respond to it with the appropriate bodily movements. Even with this focus on the brain, Galen still had room in his theories for the heart. He believed it was the seat of our "vital spirit," a vapor that traveled through the veins and arteries and powered the "animal spirit" in our brains.[2]

Localizing Function in the Brain

Over the next several centuries theories on what the brain does and how it does it proliferated. Through the careful study of patients with brain damage, scientists eventually came around to the idea that mental functions originate in the brain. It took a while, though; the idea did not really catch on until the nineteenth century. By then most scientists conceded that different areas of the brain were responsible for specific and localized functions. The next logical step was to determine which bit of brain was responsible for what.

One of the first to take a stab at making a functional map of the brain was Franz Joseph Gall, a German physician and anatomist who examined both the cream and the dregs of society in order to understand brain organization. By looking carefully at the skulls of poets, politicians, mothers, murderers, thieves, philosophers, prostitutes, and scientists (and probably anyone else willing to sit for him), he and Johann Gaspar Spurzheim created the theory of phrenology—often referred to by the slang term "bumpology." It may sound more like the name of a quirky hip-hop song than a respectable theory of science, but it was all the rage in the mid-1850s.

The theory was fairly simple. Gall postulated that the brain had distinct parts, regions that he referred to as "organs," where qualities like wit, memory, courage, destructiveness, mirthfulness, and even metaphysical ability were localized. The bigger the organ (the size of which directly correlated with one's propensity for the associated quality), the more it would protrude, pushing out the skull. If you lacked a certain faculty, there might even be a small hollow in the cranium to show this. According to the theory, a good student should have buggy, protuberant eyes to make room for the larger memory and language organs behind them. A violent criminal would have a series of distinctive bulges directly behind the ears. And a venerable man should show some size right on the top of the crown—perhaps on which to better rest a halo.

In a practical manual of phrenology originally published in 1885, the president of the American Institute of Phrenology, Nelson Sizer, and his colleague H. S. Drayton wrote, "Phrenology teaches that every sentiment, every element of taste and aversion, of hope and fear, of love and hatred, as well as the intellectual faculties and memory, have their special seats in some part of the brain."[3] By simply running a deft hand over the topography of the cranium, you could learn all you would ever need to know about a person.

This theory included love. Place the palms of your hands over your ears. Reach back with your fingertips and feel the back of your head. In Gall's day this part of the skull, the occiput, was considered the location of a person's "domestic propensities," that is, the distinctive brain organs

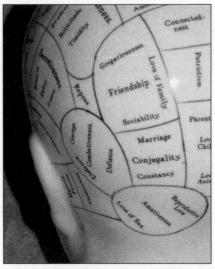

The back of a phrenology model, highlighting the areas thought to be responsible for "domestic propensities" such as amativeness, conjugality, and friendship. *Photo by the author.*

responsible for amativeness (physical love), conjugal love, parental love, friendship, and inhabitiveness (a love of home). Phrenologists believed a well-rounded person should possess a smooth, elongated, and broad occiput—a balanced grouping of these various love-related organs. Sizer and Drayton explain:

> *[Individuals] may be intellectually wise; they may be technically honest as to property and social rights, but if they lack Parental Love they will not want children; if they lack Conjugal Love they will not want marriage. If they have strong Amativeness, they may desire society through action of that faculty. . . . Free love animals and free love men lack something which does them no credit. Conjugal Love, the special, life-long, individual and exclusive mating, is human, honorable, natural, and the only sound philosophy of sexual mating.*

Basically, if you were looking for love in the Victorian age, you had better hope to have your chaperone distracted long enough for you to get a good feel of a potential mate's occiput before committing to anything permanent. According to phrenologists, one misplaced lump or chasm could make all the difference to your future happiness.

No matter that the spot that Gall and Spurzheim denoted as the seat of that dratted amativeness is not even adjacent to the brain. It lies next to some sinuses and veins, a good distance away from any gray matter. This little inaccuracy illustrates one of the many problems with phrenology. While Gall and Spurzheim's basic idea that bits of brain underlie different functions fits right in with today's neuroscientific theories, their focus on both expansive, ill-defined traits (just what might an organ for "firmness" or "sublimity" refer to?) and the exterior as opposed to the interior of the skull means it is not a scientifically sound theory.

Although phrenology eventually crashed and burned in both the popular and scientific dominions, the concept of function localization managed to stick around. For the next two centuries scientists focused on trying to pinpoint areas of the brain associated with memory, language, attention, and movement by observing patients with brain damage and animal models as well as using a variety of electrophysiological techniques. Even these "simple" concepts were difficult to study in the brain. Something like love, with its associated erotic, cognitive, and goal-directed behaviors, was too much for many researchers to even consider studying, especially since it was unclear what *love* might be. Was it an emotion, like sadness or fear? A drive, like hunger or thirst? A human construct to justify sex that had no basis in biology whatsoever? No one knew for sure. The fact that scientists could not pin it down made love seem impervious to serious scientific inquiry—and put it on the research back burner for more than a century.

Scanning Love

In the late twentieth century advances in technology enabled researchers to transcend one of phrenology's biggest failings. Neuroimaging techniques like computerized axial tomography (CAT) scanning, single photon emission computed tomography (SPECT), and positron emission tomography (PET) allowed scientists to look inside the skull and observe the living, working brain instead of relying on cranial bumps, autopsy specimens, or animals. These new approaches provided more detailed analysis of localized brain function. But it was not until the early 1990s, when a new technique called functional magnetic resonance imaging (fMRI) hit the scene, that neuroscientists could look deeper and attempt to localize something as shambolic as love.

How does fMRI work? It is all about the blood flow. Like all organs in the body, the brain needs blood in order to work. Even the smallest area of neural activity is accompanied by an influx of oxygenated blood. The brain uses that oxygen to facilitate function and then sends it on its way. This point is key for measuring activation using the fMRI. Oxygenated blood has different magnetic properties from deoxygenated blood. A large spinning magnet in the fMRI can track where the blood goes and how that blood flow changes over time. By following that track, neuroscientists can see what areas of the brain are active in response to different stimuli and tasks.

Who first thought to try to map the neural correlates of love is up for debate. In the late 1990s both Andreas Bartels, a newly minted PhD at University College London, and Helen Fisher, that savvy evolutionary anthropologist from Rutgers University, believed there must be some neurobiological evidence of love in the brain. It just had to be tested. But since the days of phrenology, no one had really tried.

Fisher had been studying the anthropological aspects of human sexuality, monogamy, and love for decades. Her research convinced her that romantic love was not an emotion, as so many others had postulated, but an actual physical drive like thirst or hunger. "It just came to my mind that romantic love was a very powerful *physical* experience," she said when we discussed her first study about the brain and love. "And that if I looked at brain functions, I might be able to establish what was going on in the brain when someone falls in love."

After speaking about this idea at several conferences, Fisher linked up with Lucy Brown, a neuroanatomist at the Albert Einstein College of Medicine; and Arthur Aron, a social neuroscientist from the State University of New York at Stony Brook. The group hypothesized that there were three distinct brain systems for love: one for sexuality and sexual behavior, a second for feelings of deep attachment, and a third for romantic love.[4] "It seemed to me that there were three basic feelings that go with love, and all others sort of derive from them," said Fisher, her voice thoughtful and calm. "I thought romantic love would be the easiest one to measure. It's such a dramatic feeling, with strong elements of focus, energy, and motivation."

Fisher is right; love's symptoms are both physical and dramatic. Those afflicted by it are often distracted, constantly daydreaming about their intended. They are emotional, too, prone to exaggerated laughter, tears, and fears. One cannot forget the actual physical manifestations of this condition: new lovers may feel butterflies in the stomach and experience an elevated heart rate, sweaty palms, and weakness in the knees. These individuals may exhibit signs of anxiety, loss of appetite, and slight obsessive-compulsive tendencies, as well as poor decision-making ability. They may sneak out of the house at night, be consistently late for work, drop out of college, or move to a new city for no other reason than to be with their lover. In my case, I blame my irrational purchase of an obnoxiously purple couch solely on the fact that I was head over heels with the object of my affection while shopping for it. Fisher was strongly convinced there was a biological explanation underlying these extreme changes in behavior. She and her colleagues set out to find it.

But before the group completed their testing, Bartels and his former advisor, Semir Zeki, a professor of neuroaesthetics (a department of Zeki's own design focusing on the neural basis of aesthetics) at University College London, published their own study comparing passionate love and friendship in the fMRI scanner in the November 2000 issue of *Neuroreport*. Zeki was inspired by mentions of love in art. How often, in the throes of passionate love, have you thought that Rumi poem or Elvis Presley song must have been written for you? How many times have you looked at a painting and thought it represented a deep and true feeling you experienced? Think about all the descriptions of love that are out there—I mentioned quite a few in the introduction. Zeki believed that if the feeling could be captured and understood in these artistic contexts (or as a tire iron, as the case may be), there must be something common about love and other emotions inside each of us. Something that is an intrinsic part of our makeup, passed from generation to generation, that allows us to share similar emotional experiences. Otherwise we would not be able to recognize or connect with so many artistic expressions of love. He makes a compelling point.

There is no question that the right visual image can elicit an emotional response. I'm only slightly embarrassed to admit that I am a reliable sucker for cute baby photos, AT&T commercials, and romantic comedy film trailers. And I am not the only one. The right picture, smell, or song can evoke commanding memories, along with any emotion behind them. Banking on that kind of power, Bartels and Zeki scanned seventeen folks who declared themselves to be passionately in love, eleven women and six men, while viewing facial photos of their significant other as well as photos of three other friends who shared the same sex as their beloved. The researchers instructed the study participants to simply look at each photo, think of the person in the photo, and relax. When they compared brain activation while viewing a lover and viewing a friend, they found two areas of the brain that reacted strongly: the left middle insula, an area implicated in emotion, self-awareness, and interpersonal relationships; and the anterior cingulate cortex, linked to reward anticipation, decision making, and emotion. And when they slightly lowered the threshold of activation, they also saw elevated blood flow in the hippocampus, the caudate nucleus, and the putamen, all areas involved with learning and memory, as well as the cerebellum, involved with the fine-tuning of motor control. It was a unique pattern, they argued, that could not be accounted for by anything but passionate love (though, Zeki quipped later, the brain activation pattern did look an awful lot like what you see in a brain after a hit of cocaine). What it meant, exactly, required further study.[5]

A Separate System for Romantic Love

Fisher, Brown, and Aron used a similar photo viewing task in their fMRI study. Instead of asking participants just to think of their beloved, however, they also asked participants to think about specific events relating to the person, such as a romantic dinner or a recent trip to the beach—any situation in which they were together, excluding those of a naughty nature. Despite this slight change in focus, their results showed quite a bit of overlap with the Bartels and Zeki study. Fisher took these

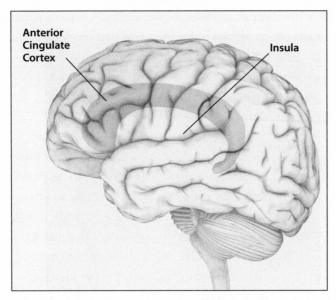

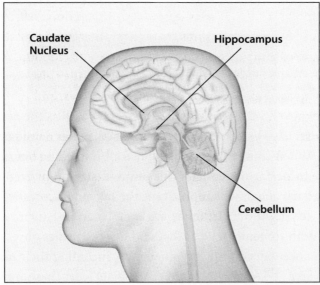

Areas activated in the original neuroimaging study of love by Semir Zeki and Andreas Bartels. This study also found a significant decrease in prefrontal cortex activation. *Illustrations by Dorling Kindersley.*

findings and suggested a coherent theory of love: three distinct yet intersecting brain systems that correspond to sex, romantic love, and long-term attachment (like a mother-child bond or the comfortable relationship you might see in a couple who have been married for sixty years). These three separate systems, she argued, could cover all facets of love: romantic, parental, filial, platonic, and that old bugger, lust.

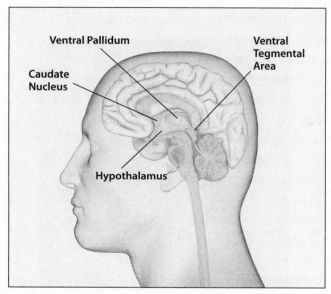

Helen Fisher and her colleagues hypothesize that there are three distinct systems involved in love: the hypothalamus for lust, the caudate nucleus and ventral tegmental area for romantic love, and the ventral pallidum for attachment. *Illustration by Dorling Kindersley.*

Scientists have long known that the seat of the sex drive is the hypothalamus. When it is removed, folks lose all interest in sex, as well as the ability to perform sexually. This almond-size brain area is linked to the pituitary gland, which produces the hormones necessary to fuel the desire to "get it on." Humans are more than just their sex drives, however. With romantic love, Fisher and her colleagues observed brain activity in areas outside the hypothalamus, including the right ventral tegmental area (VTA) and the right caudate nucleus. These are both part of the basal ganglia, a brain area connected to both the cerebral cortex and the brain stem. The basal ganglia, along with the hypothalamus and amygdala, is implicated in reward processing and learning. It's a little like bribery: when we experience something that feels good, such as satiating our hunger, having a sexy romp, or spending time with the object of our affections, these areas of the brain give us a little extra boost to encourage us to do it again. If we are talking about deep emotional attachment, the ventral pallidum, a different part of the basal ganglia circuitry, is activated. All these areas are very sensitive to the neurochemicals dopamine, oxytocin, and vasopressin, which are thought to be pleasure-inducing and critical to forming pair-bonds in socially monogamous animals (to be discussed in more detail in chapter 3). But they each work a little differently.[6]

The two regions that seemed most important to romantic love in the Fisher study were the caudate nucleus and the VTA. These areas reside in what is called the "reptilian brain"—a cluster of subcortical regions near the brain stem that have existed since before we evolved to walk upright—and are strongly implicated in both reward processing and euphoric feelings. They are also part of an important dopamine-fueled circuit called the mesocortical limbic system, a pathway critical to motivational systems; unsurprisingly it's a circuit that has been implicated in addiction. These study results led Fisher, Aron, and Brown to conclude that romantic love is not an emotion, but a drive. According to Brown, "Love is there to help fuel reproduction, to help us psychologically by connecting with others. It is distinct, yet related to lust and attachment."

Think of it this way: Lust may be the simplest of the three hypothesized systems, an almost reflex-like process that keeps us getting busy. Certainly if it were a more involved process, we would not find ourselves so interested in individuals like Pamela Anderson in all her *Baywatch* glory or, like one of my girlfriends who is too embarrassed to be named, totally hot for the *Jersey Shore*'s resident Lothario, Mike "The Situation" Sorrentino, right? At the same time we also have a system for attachment. Feeling connected to someone is a rewarding behavior, hence that ventral pallidum activation; it is nice to have someone to come home to, even if you are no longer inclined to jump his or her bones 24/7. Somewhere in the middle is the romantic love system, connected to both lust and attachment. It hits on areas involved in attachment and lust, as well as those implicated in reward processing and learning. It is no surprise that romantic love feels good and helps us to bond with another person (and consequently promotes procreation).

"These brain systems often work together, but I think it's fair to say they often don't work together too," Fisher told me when I asked whether these three systems overlapped in other ways. "One might feel deep attachment for one partner, be in romantic love with another partner, and then be sexually attracted to many others. There's overlap, but like a kaleidoscope, the patterns are different." It is also possible that these systems work on a bit of a continuum: one's physical attraction for a person can develop over time into romantic love and then into a deep-seated attachment. It might even work the other way: a good friend to whom you are deeply attached may one day, inexplicably, seem physically irresistible. A quick flick of the wrist, a change in circumstance or age, and the kaleidoscope may offer you a completely different configuration.

Love Also Deactivates

At times active brain areas are not the only ones that are important to understanding function; deactivated areas can tell us something too. These neuroimaging studies have also shown decreased activation, which may be related to decreased function, in certain brain areas. The frontal lobe, the parietal lobe, and the amygdala show diminished blood flow in love and attachment. They say love is blind, and if you've ever been in love with the wrong person, you know it to be true. Zeki argues that the lack of blood flow to these areas suggests reduced function in judgment, decision making, and the assessment of social situations.

You Were Always on My Mind

A picture may be worth a thousand words, but you do not need explicit visual stimuli to activate the brain's romantic love system. Stephanie Ortigue, a neuroscientist at Syracuse University, noticed that people in love are very quick to make associations between the object of their affection and certain words and concepts. If a place, word, situation, or song has the slightest thing to do with their sweetheart, they will make all kinds of interesting connections. People in love cannot stop thinking or talking about their boo. This priming effect, the strength of connections between your beloved and everything related to him or her, may facilitate that kind of quick recall. Ortigue decided to take a look.

Ortigue and her colleagues scanned the brains of thirty-six women who were passionately in love while they were subliminally presented with the name of their significant other. It would seem that, even implicitly, love really has a strong hold on the passionately in love individual. Even when words were used instead of photos, many of the same subcortical reward-related brain areas fired up in this study, as in previous neuroimaging studies: the caudate nucleus, insula, and VTA. But these researchers also documented activation in higher-order brain areas, parts of the cortex involved in attention, social cognition, and self-representation: the angular gyrus, middle frontal gyrus, and superior temporal gyrus.[7]

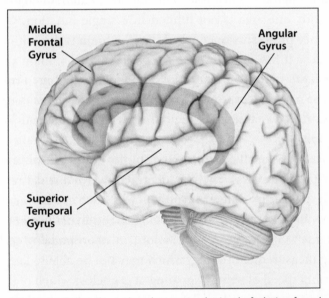

Stephanie Ortigue's study using names instead of photos found activations similar to those in the Zeki and Fisher studies. She also found higher-level activation in the angular gyrus, superior temporal gyrus, and middle frontal gyrus. *Illustration by Dorling Kindersley.*

"Our findings suggest not only that these reward systems are important in love but also more cognitive areas related to decision-making and the representation of self and body image," said Ortigue. "It's quite interesting—it suggests that love may be an extension of oneself. Or rather, people in love really put themselves into others. It changes the way we conceptualize passionate love." If my love object can change the way I internally view myself, what else might change? This research brings a whole new meaning to the line "As your lover sees you, so you are."

Putting the Pieces Together

Though Fisher postulates that romantic love is a drive—one that has been evolutionarily selected in order to motivate us to have babies and raise them as pair-bonded couples—Ortigue cautions that it is dangerous to simply classify love as a basic instinct. There are too many different brain areas implicated.

She has a point. No part of the brain area is an island; all of these regions are interconnected and send signals to and from one another. What's more, one area is not limited to a single function. One of my neuroscience professors once joked that the brain is the "ultimate recycler" because it has evolved over the past hundred million years to be superefficient. After all, it takes a lot of blood and energy to run a brain. It would be a serious waste of resources if single regions couldn't help facilitate a variety of tasks. The brain frowns on redundancies—and with good reason.

In an analysis of all neuroimaging studies done on love (a whopping six in the past ten years), Ortigue identified twelve distinct regions that were activated across different types of tasks, such as viewing photos or watching videos of a loved one or being subliminally presented with a lover's name.[8] Given current limitations in neuroimaging technology, including measurement timing, which may not be able to keep up with the lightning speed of neural signaling, it is unclear which of these areas lights up first in romantic love, let alone how and when the different areas may interact. Nor is it apparent how our subcortical brain regions, the so-called reptilian brain implicated in reward processing and euphoria, may be influencing the higher-level cognitive areas involved in attention, self-representation, and decision making and vice versa. There is still quite a bit to learn.

"When people say this or that area of the brain is activated in love, my standard reply is, 'So what?'" Ortigue said emphatically as we discussed what her analysis of love and the brain really tells us. "People think it's easy to look at these activations and know what is happening. But there's so much more beyond that. It's very dangerous to try and simplify this kind of network. There is not a single brain area at work here. We need to understand how all these areas work together before we can say anything about the nature of love."

We are not there yet—nowhere close—but headway has been made in understanding how the ventral tegmental area, caudate nucleus, putamen, and other areas work. Before I can explain how they may be interacting to get us into some of love's most blessed (and thorny) situations, I need to discuss a powerful neurotransmitter called dopamine and other chemicals that help fuel this love-related brain network.

Chapter 3

———

The Chemicals between Us

Each distinct region of the brain is made up of specialized cells called neurons. There are many of them—billions, in fact—and every few years someone attempts to get an accurate count of these bad boys. It is a daunting task, but the latest estimate tallied the number at about eighty-six billion.[1] That is a lot of cells. *A lot.*

No cell, however, works alone. In order for the brain to perform its magic, these neurons need to signal each other by releasing a variety of chemical messengers called neurotransmitters. A stimulated cell releases the messenger, which then enters the synapse, which is the minuscule space between that brain cell and its neighbors. There the neurotransmitter may be picked up by an adjacent neuron, partner up with another chemical, or float in the ether for a while before it is taken back by the initiating cell. That is an oversimplification of the process, but it hits the basics.

We often discuss the function of these various brain chemicals as if there were only one molecule traveling between two cells—just a lonely little passenger traveling along a simple, direct path. In truth a single synapse during real-time neurotransmission looks more like holiday weekend rush hour on the Cross Bronx Expressway. Hundreds, if not thousands of different neurochemicals, hormones, and proteins move around the synapse at any given moment, and Demolition Derby rules are in place; these compounds not only stimulate bordering cells but

24

they can modulate fellow travelers as well. They can transform, cleave, or even block chemicals from communicating with other cells as they float in the synapse. Or they may just hang out in the synapse until they are taken back by the releasing cell, changing which chemicals are released the next time the cell fires. They can even help fellow neurochemicals get to their destinations faster and talk to different kinds of receptors.

Here is a situation where the old adage about elephants works all too well. How do you eat an entire elephant? You can do it only one way: a single bite at a time. How else could you handle millions of neurons, trillions of synapses, and the thousands of chemicals and proteins roaming within each synapse? By observing one reaction, one neurotransmitter, one receptor at a time, scientists have uncovered several interesting chemicals related to love.

The Dope on Dopamine

At the base of the forebrain, the large frontal cortex that differentiates humans from other mammals, are the basal ganglia. Hundreds of millions of years ago, when our ancestors climbed out of the water and started spending time on dry land, they possessed brains not unlike the current incarnation of our basal ganglia, composed of the corpus striatum, a striped structure made up of the caudate nucleus, putamen, and nucleus accumbens; the pallidum, with its pale globe, the globus pallidus, and neighboring ventral pallidum; the substantia nigra, or "black substance," the region in the basal ganglia that appears darker than the rest; the ventral tegmental area; and the subthalamic nucleus, a small area adjacent to the thalamus, the region responsible for relaying sensations. As we developed larger frontal lobes, our basal ganglia maintained their formidable sway over behavior and learning by forming strong connections both to and from the prefrontal cortex and other key brain areas. The basal ganglia region is linked to almost every aspect of cognition and is also part of an important brain pathway called the mesocortical limbic system, which is involved with motivation and reward processing.

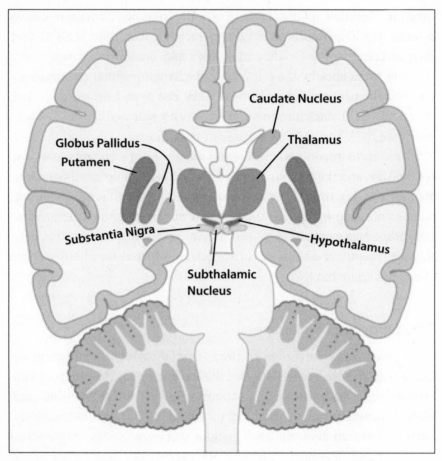

Our "reptilian brain," the major structures of the basal ganglia.
Illustration by Dorling Kindersley.

The basal ganglia are home to many different neurotransmitters, but these areas, as well as the mesocortical limbic system, are fueled primarily by the dopamine released from a group of neurons in the substantia nigra and VTA. It is this influential chemical that allows for profound changes to the brain that result in learning, memory, and movement. Dopamine is as old as the reptilian brain itself—and quite potent stuff.[2]

Dopamine has been studied extensively in individuals who have Parkinson's disease. The death of critical dopamine-producing neurons in the substantia nigra results in Parkinson's hallmark symptoms: tremor, rigidity, and dementia. Schizophrenia has been characterized as an overabundance of dopamine production in the brain. And with a hand in Tourette's syndrome, anorexia nervosa, obsessive-compulsive disorder (OCD), attention-deficit/hyperactivity disorder (ADHD), and drug addiction, this is a neurotransmitter with some serious pull.

Now think about falling in love. You may show some obsessive tendencies when it comes to your intended, not so unlike an individual with OCD. You may be distracted at work or at home, a little love-related ADHD. You may start to attribute significant meaning to minor traits you have in common with your mate: your mutual adoration for horror films and the fact that you are both addicted to pistachios, not to mention the way your competitive natures both come out during March Madness. Taking trivial connections and turning them into matters of great importance can often be seen in schizophrenic patients (like John Nash's psychotic code-breaking methods in *A Beautiful Mind*). A person in love may have impaired decision-making ability; Parkinsonian patients often have great difficulty making even the simplest choices.

Addiction? That leap is not hard to make. It is all too easy to remember those first few months of love, when you just couldn't get enough of your significant other. I am not saying love is a disease (though more than a few writers have poetically done so), nor that love is part and parcel of these disorders. Rather, when you look closer at the behavioral effects resulting from too much or too little dopamine in the basal ganglia, you can see why Helen Fisher thought it must underlie the neurobiology of love. "I read descriptions of romantic love from the last forty years so I could get a feel for what the elements were," Fisher explained. "And when I looked at those elements—the focus of love, the energy, the motivational factors—I came up with the hypothesis that dopamine had to be behind it."

She was right—many areas in the basal ganglia, particularly those that send or receive dopamine projections, lit up in response to romantic love in the fMRI scanner. Animal studies have suggested dopamine plays a remarkable role in the forming of pair-bonds too. Prairie voles are a monogamous species; more than 80 percent of these small rodents pair up with just one mate for life. The dopamine released after the initial mating session, likely the source of its pleasure, primes the brain to make that bond. When scientists gave prairie voles drugs that inhibit the production of dopamine, these über-monogamous rodents were unable to form pair-bonds after mating.[3] By contrast, prairie voles that were injected with drugs that increased the amount of dopamine in their brains formed bonds even without the precursory mating session. Dopamine matters. Not just in the positive reinforcement from sex to form the original bond but also in the continued matings that allow the animals to maintain that bond over time.[4]

How does one little chemical have such reach? It comes down to the processing of risks and rewards. Recent research into the study of reward processing and decision making has shown that dopamine plays a big role. Using animal models, scientists have learned that when a rat is making a choice that leads to an unexpected reward, like food or a positively stimulating drug, there is a notable increase in dopamine release in the basal ganglia. After being trained to expect a reward with a paired stimulus, if the rat does not receive it or gets something less than stellar (some plain old food pellets instead of a hit of cocaine, for example), the cells in the basal ganglia release less dopamine. The same kinds of effects have been observed in people during neuroimaging studies.

Michael Frank, a neuroscientist at Brown University who studies the basal ganglia, believes that this is how dopamine facilitates learning. After all, when you receive a reward, you want to figure out what you did to earn it so you can go out and do it again. Depending on how good the treat is, you might even perform that task a few more times after that. On the other side of the coin, if a certain behavior results in punishment or a negative consequence, you'll want to avoid that situation in the future. "The basal ganglia circuit is perfectly structured to help execute reward-based learning," Frank told me. "It allows you to learn about positive and negative outcomes of your choices. When you have that increase in dopamine, you are more likely to pursue the same rewarding outcome in the future."

In prairie voles that dopamine flood that results from mating primes the brain for creating a lasting pair-bond with a member of the opposite sex. Without that dopamine rush, the bond might never happen. This is true in humans too—and not just with sex. Any social interaction promotes the release of dopamine. Laughing with friends, cuddling with your kids, holding hands with your significant other, even petting your dog—simply being with others is a reward in its own right.

The basal ganglia facilitate more than just pleasurable feelings, though. This collection of regions sends signals to areas that help filter out unnecessary information so you can focus on the most important aspects of the task at hand. They enable you to form a memory of an experience to retrieve when you later find yourself in a similar situation, and they help you organize your movements so you can act in response to a stimulus. The basal ganglia are a critical circuit that informs all manner of cognitive activities, and dopamine fuels it, that is, when we are talking about any kind of learning that stems from a reward—love included.

Neuroscientists are very interested in how we assess both reward and risk, and how this valuation plays out in many types of behaviors. This means they're fascinated by sex and love; after all, is there any greater reward (or risk, for that matter) than finding your one true love (or, in his absence, getting a little something-something while you wait for him to show up)? Is there any other stimulus with such profound and far-reaching effects on behavior? Very few, I'd wager.

Oxytocin, Oxytocin Everywhere

Just above the brain stem, below the thalamus and basal ganglia, is the hypothalamus. This little almond-shaped brain region has the critical function of mediating communication between the brain and the endocrine system. In other words, this area has a say on which hormones, those great behavioral primers, are released into the bloodstream and into the brain. It is directly linked to the pituitary gland, the brain's so-called master gland and secretor of hormones into the body, as well as to the thalamus, the nucleus accumbens, and the ventral tegmental area. Specialized areas within the hypothalamus, like the paraventricular nucleus (PVN) and the supraoptic nucleus (SON), do some of that mediating with the release of oxytocin.

Oxytocin is a bit of a wonder compound. Some even go so far as to say it is a little magical. Technically it is a neuropeptide, a small protein-like molecule that can act as a neurotransmitter, stimulating and inhibiting neurons in the brain. When released into the bloodstream, it has been implicated in such behaviors as milk letdown in nursing mothers, stimulation of uterine contractions during labor, and estrus

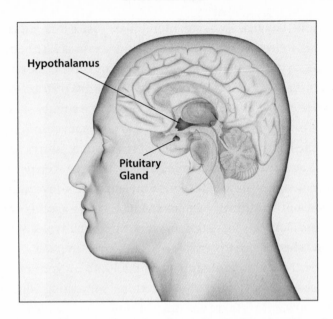

The hypothalamus mediates the release of oxytocin and vasopressin into both the body and the brain. The pituitary gland, the almond-shaped gland beneath the hypothalamus, releases other important hormones. *Illustration by Dorling Kindersley.*

cycle modulation. Pitocin, a synthetic version of oxytocin, is used to help speed up labor during the birth process. Because of these maternal effects, researchers initially believed oxytocin to be a specifically female chemical, but the PVN and SON also send oxytocin directly into the brain's cortex—in both males and females. In response to touch, sex, and social bonds, oxytocin stimulates cells in the ventral tegmental area. Once stimulated, these cells then initiate that lovely and agreeable dopamine cascade in the basal ganglia.

"Oxytocin is a very important chemical. The brain is provided with its very own oxytocin system, that is very often activated in parallel to the circulatory oxytocin system," explained Kerstin Uvnäs-Moberg, an oxytocin expert and the co-organizer of the original Wenner-Gren Symposium. She continues her work with oxytocin today at the Swedish University of Agricultural Sciences and kindly offered to discuss the wonders of oxytocin with me over a crackly telephone line.

If you read the science news, it is hard to avoid oxytocin these days. Researchers across the globe are finding new ways in which oxytocin levels are positively influenced. Activities like cuddling your baby skin-to-skin, hanging out with your dog, getting a massage, having orgasms, spending some time on the social networking site Twitter, or even experiencing simple eye contact with another person can amp up your oxytocin level. High levels of this neuropeptide have been associated with the ability to better read social nuance, increased trust and emotion, strong social bonds, and improved mood. On the other hand, if your oxytocin level is too low you may find it difficult to recognize faces, properly nurture your kids, or avoid cravings for salty and sweet foods. Social bonds, parent-child relationships, sexual behaviors, trust—oxytocin seems to play a role in all these complex behaviors and more. It is also now considered part of a potential treatment for autism, schizophrenia, and depression. Some "experts" have even proposed that it may be worthwhile to snort a little oxytocin to help bolster your mood and sociability before a date or job interview, though the majority of scientific researchers who study oxytocin would strongly argue against such a practice.

This is a chemical with extensive reach, yet its effects are quite subtle. When neuroscientists at Emory University used knockout techniques to remove the genes that produce oxytocin in mice, they thought they might find some changes in maternal behavior. Instead they learned that the lack of oxytocin in the medial amygdala region, a brain area rich in oxytocin receptors, made it so the animals could not recognize other mice. Their senses of smell and space were intact, as were their memories, yet knockout males could not recognize previously introduced females after thirty minutes apart, something normal mice can do without a problem. When the researchers injected these animals with oxytocin before the initial meeting, social recognition was restored.[5] In comparison, social recognition was also impaired when an oxytocin antagonist, a drug that stopped oxytocin from connecting with receptors, was injected into the medial amygdala of normal animals. It appears that the lack of oxytocin during the introduction interfered with encoding the appropriate social information about the female. This finding correlates with human studies. Scientists have shown that amygdala damage leads to problems with facial recognition, reading social cues, and expressing emotion. It makes sense. How can you form a bond, lasting or otherwise, with someone you do not even recognize at a second meeting?

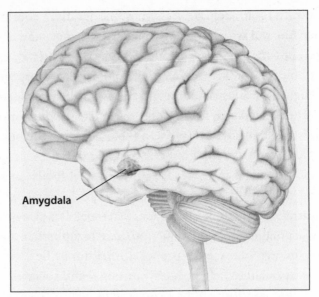

Lack of oxytocin in the amygdala has been linked to social recognition deficits. *Illustration by Dorling Kindersley.*

Uvnäs-Moberg theorizes that this neuropeptide is at the heart of a psychophysiological system that directly opposes the "fight or flight" phenomenon, which is an acute stress response mediated by activation in the hypothalamus, brain stem, and amygdala in response to situations that threaten survival. She proposes an analogous "calm and connection" response. Instead of accelerated heart rate, anxiety, and elevated glucose levels—the hallmarks of "fight or flight"—oxytocin helps mediate a contrasting circuit that results in relaxation, decreased heart rate, and an overall feeling of calm.

When you perceive a threat, you have to be prepared to act. Physiologically your body has to be at the ready to brawl or get the heck out of Dodge, and the response must be immediate. A physiological state of calm and connection does not require the same kind of speed. In the absence of danger you can take your time. Repeated administration of oxytocin has been shown to induce long-lasting effects to areas of the brain involved in social bonding, sexual behavior, and parental attachment. Over time this calm-and-connected state results not only in a feeling of high, as you might experience while in the throes of new love, but also in a general sense of relaxation and peace. That sense of well-being may have health benefits too. Many studies have suggested that a calm demeanor and strong social relationships can help people weather stressful situations, fight disease, and even lengthen life. Add up these varying lines of research and you will find, Uvnäs-Moberg argues, a strong evolutionary advantage to a system antithetical to fight-or-flight. What's more, she said, given its effects on the nervous system, oxytocin is qualified to be that agent of calm and connection in the brain.[6]

Vivacious Vasopressin

Like oxytocin, vasopressin (more formally referred to as arginine vaso- pressin) is a small neuropeptide also released by the PVN and SON in the hypothalamus. It is very much like oxytocin; in fact the genes for both chemicals resides on the same chromosome. Their similarities are such that some scientists hypothesize the two may have evolved from a common compound a few hundred million years ago.[7]

Vasopressin is involved with the regulation of blood pressure— hence *vaso*, indicating blood vessels. It is also implicated in kidney function, cell homeostasis, and glucose regulation. In the behavioral realm it is linked to attention, learning, memory, and aggression. Like oxytocin, it has a role in forming pair-bonds. It is a busy little bee of a neuropeptide.

If oxytocin was once thought of as a girl chemical, then vasopres- sin was its boy correlate; like oxytocin, however, vasopressin is pres- ent in both sexes. In prairie vole studies, infusions of vasopressin have been shown to increase territorial behaviors. Within a day of mating and bonding with a female, males will show heightened aggression; this newlywed will kill in order to make sure his lady, fertile and ready, will not be impregnated by a fellow suitor. Though slower to develop, aggres- sion in females is also increased after they pair-bond. Other girls had better not even think of coming around their man. Not if they are keen on surviving.

Vasopressin is a difficult chemical to study, mainly because of its vast and various responsibilities throughout the body. This chemical is not unleashed just in the brain in response to sex but is also released in other parts of the body. Given its importance to heart and kidney function, it is not an element you want to knock out completely in experiments, since you will end up with sickly, and probably dead, animals. To add another level of complexity, vasopressin can work with and against oxytocin. Each of these chemicals can bond with the other's receptors, sometimes facilitating and sometimes blocking action. Researchers are still trying to tease apart their respective roles in the formation and maintenance of pair-bonds.

Some Other Key Players

Several other neurotransmitters have been implicated in love. Glutamate is a substance critical to learning and memory. Gamma-aminobutyric acid is a compound best known for inhibition (the damping down of overexcited neurons), as well as its effects on alertness and arousal. Both have been shown to be involved in reward pathways, but it is unclear how they interact with each other to result in love's many behaviors. Studies are ongoing.[8]

Serotonin, a neurotransmitter you may recognize for its importance in the treatment of depression, is also relevant when it comes to love. Although it is known for its mood-enhancing effects, serotonin levels actually drop in the early stages of love. This is a conundrum. Love makes you feel good, right? It enhances mood. Shouldn't it make serotonin levels increase? But love also releases high amounts of dopamine. When dopamine goes up, serotonin goes down. It's one of the ways the brain balances itself. Serotonin is often referred to as a brake, a way to stanch the dopamine flood so you aren't always in thrall to those good feelings.

But love does more than just make you feel good. It also makes you a little obsessive—and obsession can be stressful. An early molecular study of love looked at the serotonin transporter, a small protein that helps serotonin get from place to place in the synapse. Donatella Marazziti, an Italian neuroscientist at the University of Pisa who studies love, compared the number of binding sites for the serotonin transporter protein in individuals who had recently fallen in love, individuals suffering from OCD, and normal, single controls. Marazziti and her colleagues found that the number of sites was significantly lower in both

OCD patients and those in love. What does this mean? Basically similar serotonin neurochemistry may have something to do with obsessive thoughts in both OCD and love. That same lack of serotonin that results in an OCD patient's believing that touching a door five times upon entering can guarantee safety may also be behind the way you constantly, compulsively think about a new squeeze when you are in the honeymoon phase of a relationship.[9]

You may have noticed that I have not mentioned estrogen or testosterone yet. One might think these gonadal hormones would play a strong role in love; after all, common knowledge suggests that all our love-related behaviors are driven by them. Perhaps you think estrogen is behind the desire to connect with your one true love. Though present in both sexes, it is often referred to as a female hormone and is characterized as the molecule that makes us vulnerable and open to others. When Marazziti and her colleagues at the University of Pisa measured hormone levels in individuals passionately in love, singletons, and those who had been in a committed relationship for some time, they saw no differences in estradiol, a form of estrogen, or progesterone, another so-called girlie hormone. What they did find was a heightened level of testosterone in women who were passionately in love and decreased levels of testosterone and follicle-stimulating hormone (FSH), a hormone linked to reproduction, in men. They also saw increased levels of cortisol, a hormone released in response to stress. When these passionately in love folks were retested one to two years later, all hormonal differences between the groups disappeared.

Why these levels changed in this manner is up for interpretation. Marazziti suggests the heightened cortisol levels represent arousal but also a certain amount of stress in the early days of a relationship. The idea fits with several animal studies demonstrating that being a little stressed out actually promotes social interaction and attachment. In animals a perceived threat is an incentive to band together. With a pal or pair-bonded mate around to help forage for food, take care of the young, and watch your back, the probability of survival increases dramatically.

It is also possible that these raised cortisol levels simply represent plain old garden-variety stress. You may be head over heels, but can you be certain your intended feels the same way? Anyone who is in the early days of a relationship can relate to overthinking the whole loves-me, loves-me-not thing. It can be a source of some consternation. As I said, why Marazziti saw these enhanced levels of cortisol is up for debate. It could be due to some vestigial stress response to a hundred-million-year-old attachment system, new relationship concerns, or something else altogether. We may never know.

As to Marazziti's findings regarding testosterone and FSH, she has no hard-and-fast answers. In a paper published in *Psychoneuroendocrinology* about the study, she wrote, "All subjects presented this finding, as if falling in love tended temporarily to eliminate some differences between the sexes, or to soften some male features in men and, in parallel, to increase them in women. It is tempting to link the changes in testosterone levels to changes in behaviors, sexual attributes or, perhaps, aggressive traits which move in different directions in the two sexes, however, apart from some anecdotal evidence, we have no data substantiating this which would justify further research." So we could say it makes women more open to sex and men more open to cuddling, or some other meeting of the proverbial minds between the two genders, but the truth is, we just don't know more than the fact that these levels change—and only in the early stages of romantic love. By the time these individuals get into a stable, long-term relationship, the levels for both testosterone and FSH return to normal.[10]

There is more to neurochemistry than neurotransmitters and hormones, however. Another group of Italian neuroscientists looked at the levels of neurotrophins circulating in the blood in those newly in love. Neurotrophins, also called growth factors, are simple proteins involved in synaptic plasticity, or the ability of the connections between neurons to change. Some neuroscientists fondly refer to them as "brain fertilizer" because these proteins help neurons to develop, specialize, grow larger, and live longer. They are also implicated in strengthening neurons to promote learning and memory processes. As described, the brain goes through a lot of changes when you fall in love: it focuses your attention on your intended, increases your good feelings and energy, and makes you a bit obsessive. If that love is returned, it may lead to a lasting bond. Those are some pretty significant changes involving many different brain areas and neurochemicals. Synaptic plasticity is required to make those changes happen. Without it, and the neurotrophins that facilitate it, our brains would remain forever unchanged, unable to learn or grow.

Considering that learning and memory-related changes in the brain have been linked to increased levels of neurotrophins, Enzo Emanuele, a molecular biologist, wondered if love—which itself involves quite a bit of learning, memory, and brain changes—would show the same kind of pattern. He and his colleagues measured levels of four different neurotrophins in individuals who were passionately in love, single folks, and those in long-term relationships who no longer felt all that passionate about their partner. The group found that one factor, nerve growth factor (NGF), was significantly higher in those who were romantically afflicted. Furthermore the level of NGF was positively correlated with the intensity of love, as measured by a standardized scale: the more in love a participant reported to be, the higher the level of NGF. As in the hormone study, Emanuele retested participants who remained in their relationship twelve to twenty-four months later. Just as was observed in the hormone study, time was all it took for NGF levels to revert back to normal, the same levels seen in the two control groups.[11] It is clear that something is going on in the early stages of love, and NGF may have a role in modulating other brain chemicals, mediating some of the brain and behavioral changes we see in new love.[12] The truth is, however, that we just don't know what exactly the NGF might be "fertilizing," or how it relates to the feelings and behaviors of love.

A Matter of Reception

It is often said that it is better to give than to receive. But giving and receiving are equally important when you're talking about synapses and neurotransmission. No discussion of love-related neurochemicals is complete without a quick look at their receptors. For any neurotransmitter—any neurochemical, actually—there may be a handful of receptor types that result in different cell actions. One kind of molecule can often also pair up with another chemical's primary receptor. It is not enough to have a certain level of a particular neurochemical;

you must have the right kind of receptors available, in the appropriate numbers, for it to be taken up by a receiving cell and do its thing. Currently researchers are studying a variety of different receptors related to all these neurochemicals, particularly in the study of monogamy. Different receptors, including a variety of vasopressin and dopamine receptors, have something to say about the strength and longevity of a given pair-bond. I will talk more about some of these specific receptors and their roles in the following chapters.

It's All about Chemistry

There are a few advice books out there claiming that lasting love is all about proper brain chemistry. To be able to connect with the right person, you need the "right" level of testosterone or oxytocin. Just follow these rules, change your diet, and you can restore your body's natural balance—or, if that is too much trouble, simply buy a line of related supplements from the author's online store for optimal love neurochemistry.[13] While it is clear that a balanced diet and regular exercise are important to good health, including brain health, there is no evidence of a single *right* chemical balance in the brain. There are kinks to neurotransmitter systems that may lead to disorders like depression, OCD, and ADHD, but in so-called normal populations there is incredible variability in these systems. Even free of disorders, our brains and their innate neurochemistry remain our own, as quirky and unique as any of our other traits and features. Current neuroscience studies offer no indication that love requires a specific level of any one neurochemical or any combination of them.

All these brain chemicals—dopamine, oxytocin, vasopressin, serotonin, and others—are mixed up in the cocktail we call love. Shake it up and—voilà!—we see a variety of different brain areas light up as well as a continuum of love-related behaviors. As any good mixologist will tell you, no matter how skilled you are, and no matter the quality of your tools or ingredients, no two cocktails will ever turn out exactly the same. Some bartenders spend years trying to re-create that one perfect drink. But they try in vain.

Alas, you find the same phenomenon in the brain. No two people are going to have exactly the same brain chemistry. Even with the same partner, no two people will have identical experiences in love. So while neuroscientists are gaining ground in how these various neurotransmitters and chemicals mix together to create complex behaviors, including that four-letter-word that begins with *l*, there is still plenty of ground to cover.

Chapter 4

Epigenetics
(or It Is All My Mother's Fault)

When I told my mother I was subtitling a chapter "It Is All My Mother's Fault," she laughed and said, "It's the cheeseburgers!" This is an old family joke. While she was pregnant with me, the only thing my mother could handle was cheeseburgers from a popular fast-food chain. Having just moved to the greater Chicago-land area, she and my father ate most meals out while they waited for their new house to be finished. Each night my mother sat stoically, trying not to retch, through whatever fabulous dinner my father ordered at nice restaurants across the city. But invariably, no matter how hard she tried, she still ended up waiting for him in the ladies' room or in the car because she was so put off by the food. As soon as he paid the bill, my mother demanded to be taken to said fast-food chain's drive-thru window. There she would order and then gobble up two or even three cheeseburgers. They were the only thing that appealed to her—and the only thing she could keep down. Ever since, any bad behavior on my part, any lip or misdeed, has been explained away later with the simple sentence, "It's the cheeseburgers!" (Of course, my successes are often described thus too.)

Little did we know that scientists were discovering that the food moms (and perhaps dads too) eat while babies are in utero can actually make changes to those babies' genomes—at least, to the way individual genes may be expressed in the body. And the way parents behave after those babies are born can also have profound effects on gene expression.

The sum of these behaviors can affect love and parenting style over their offspring's entire life span.

Behold the power of the cheeseburgers! They may very well have changed the way I have grown, the way I eat as an adult, the way I love, and even the way I parent. Who knew?

Genes and Behavior

The idea that genes are altered by the environment is counterintuitive at best. Ever since that little double helix called DNA was discovered, we have been taught that our genes are our destiny, inviolate and unbending. Our individual differences in body type, personality, intelligence, behavior, and everything else could all be accounted for by the mix of various genes we received from our parents—but the individual genes themselves remained virtually the same. Sure, evolution may curve a beak or web some toes or make other tweaks here and there, but those mutations were slow in coming, occurring over millions and millions of years. Direct damage to genes, through disease or environmental mishap, might also knock a few nucleotides out of the mix or cause repetition of certain sequences—but again, this was the exception to the rule of fixed genes. And with the start of the Human Genome Project in 1990, it seemed certain we'd soon have the key, in the form of a handy-dandy genomic map, to understanding both behavior and disease. The consistent message from the scientific community was that our DNA was the great decider, directing growth, development, disease, and ultimately behavior. Once we had mapped the human genome, we'd have the primer to help us understand it all.

There are a few problems with this idea of genetic determinism. One of the biggest? Monozygotic (identical) twins. Have you noticed that sometimes these folks aren't quite so identical? There is often some unique physical trait, a well-placed mole or perhaps a slightly different hairline, to help others tell them apart physically. More important, one of the pair may develop a disorder like asthma or schizophrenia, while the other remains untouched by the disease. And in terms of personality, no matter how close a pair of twins may be, they have their own distinct character. It is astonishing to think that these differences can occur even when you have virtually identical genomes. How does that work exactly?

It should also be mentioned that genetic studies in the social behavior realm haven't been quite as satisfying as some had hoped. Twin studies have looked for genes underlying traits like altruism, trust, and fidelity, but they have been unable to pinpoint the exact gene, or genes, involved in those qualities.[1] What those studies have been able to offer is a basic estimate of heritability. Although that is helpful, it's not all that specific.

In the past decade there has been a rise in genome-wide association studies, in which researchers compare the genetic idiosyncrasies of thousands of individuals correlated with a particular behavior (usually using survey data). But again, except in rare disorders, most behaviors, like diseases, involve multiple genes. Scientists using these methods can assess probabilities but offer no definite answers regarding an individual gene's role in something as complex as love.

We often talk about genes as if they are little gods, directing all manner of disease and behavior, especially when it comes to evolution. You'll even hear some researchers talking about what genes *want*. They want to propagate themselves, to select for the best traits, to endure, and so on. In truth genes are simply little chains of nucleotides that offer cells an instruction manual on the production of small protein molecules. They have no free will, no grand plan for you or the universe. They are simple biological machinery that help cells do their thing.

Though these proteins are critical to brain function, that Demolition Derby we call synaptic activity, they do not act in a vacuum. It is often said the brain is behavior and that the brain is built on genes. But there is no behavior without a stimulus of sorts to prompt it. Clearly there is a crucial interaction here: the biological products of our genes and our environment mix together to result in a particular behavior. As they say, context is everything.

So now there is a new neurobiological focus on the epigenome. The word comes from the Greek; the prefix *epi* means "over" or "above." The epigenome is a mechanism of gene expression that lies above the genome itself. The study of these environmentally induced chemical changes to gene expression (which takes place without mutation or change to the nucleotide sequence of DNA) is called epigenetics. It is a robust phenomenon; the epigenome is often passed down several generations along with the genes themselves. And it is turning out to be a great game changer for the neuroscience world in the understanding of learning, memory, and behavior.

Say What?

Take a good look at your computer. It is a physical object composed of hardware, such as a hard drive and a monitor. You probably also have some software on that computer: an operating system, a good word processing program, and, with luck, a good Solitaire or Tetris game to help you procrastinate. The hardware and software are separate elements, usually designed and built by different companies. To do something useful, say write a letter or beat your kid's top score on the video game du jour, you need to have both working together. The software is basically directing your hardware's action. To you, the user, this partnership is seamless. You do not think about your software being the program and your hardware running it. You don't have to. You are simply using your computer. It is only when one of the two breaks, making it impossible for you to beat that stupid game that has been invading your dreams, that their separateness becomes apparent.

It is not so different in the field of epigenetics. "The genome is comparable to hardware. And the epigenome to the software," says Randy Jirtle, director of the Epigenetics and Imprinting Laboratory at Duke University. "It's a good analogy for understanding how it works."

Simply stated, epigenetics is the way that life experiences, your parents' or your own, can actually mark up your DNA. By leaving the biological equivalent of highlights, ticks, and margin notes on individual and group genes, epigenetics can change their expression. In fact those epigenetic changes can alter whether those genes are expressed at all— even for several generations. I think it is fair to say that some of these marks are made in pencil, easily erased by new experiences or direct treatment, while others remain present in a most durable ink, holding steadfast as those genes are passed down to subsequent generations of offspring.

If you have some recall of your eighth-grade science class, you remember that DNA is a double helix, two polymer chains of simple nucleotides called adenine, cytosine, guanine, and thymine twisted together into base pairs. This particular molecular architecture provides the genetic code, the blueprint that will direct the construction and function of every cell in your body. But it does not act in isolation. Researchers have now discovered several ways different proteins can chemically adhere to DNA, or its messenger pal, ribonucleic acid (RNA), to alter not the genetic material itself but rather how that DNA is used by the cells to make all those different critical proteins.

The first and most stable of these molecular mechanisms is DNA methylation. Your experience in utero and in early life can result in an enzyme called DNA methyltransferase adding new molecules to the cytosine nucleotides in your DNA chain. This chemical change does not mutate the DNA itself. No, your genetic code remains fully intact, the order of nucleotides unchanged. Instead the methylation process adds a checkmark of sorts next to the genes it affects, typically resulting in the suppression or all-out removal of gene expression for the associated protein.

A second epigenetic phenomenon involves histone proteins and increased gene expression. In the cell DNA is wrapped around a core of alkaline proteins called histones, which have long tails that sometimes manage to stick out of their double-helix enclosure. In a process called acetylation, a different type of molecule, an acetyl group, attaches to that errant tail and creates more space between the proteins and the DNA. By doing so, it leads to a surge in gene expression. Similarly, deacetylation can occur here too. As you have likely guessed, an experience may start a chemical chain that ultimately releases an enzyme called histone deacetylase, removing those acetyl groups (and with them the space between the DNA and histone proteins), resulting in fewer proteins being made.

There is a third molecular mechanism studied in the epigenetics of behavior, one that involves microRNA. Back to eighth-grade science class: you probably have a vague recollection that messenger RNAs copy the genetic code from DNA and then travel into the cell nucleus so it can make the prescribed proteins. MicroRNAs are short RNA molecules that attach to that messenger chain and change the message just a little bit, ultimately suppressing the expression of the gene.

If you glossed over the previous three paragraphs, I don't blame you. I offered only the most basic information for a little background. For the purposes of this book, the exact molecular mechanisms underlying epigenetic changes to gene expression are of little consequence. The important take-home message is that life experience has the power to change your genetic material at the molecular level, not by mutating your genes, but by affecting the manner in which those genes express themselves— or, more specifically, by facilitating the production of more or fewer proteins by your cells. And as we learned in chapter 3, the number of those proteins can have grand consequences in how our brain cells communicate with one another, ultimately resulting in changes to our own behaviors. "This is evolution riding on the back of software," Jirtle told me. "These changes happen rapidly. It's easier to change the code in the software, or the epigenome, than to mutate the genes in the hardware. And these changes can have profound effects on our behaviors."

It Starts with a Genetic Battle

Want to see epigenetics work its magic? Consider a newborn baby. He is the product of both his parents, both contributing their own DNA when the sperm fertilized the egg and the cells grew into a fetus. He has not had much experience in the world yet. He sleeps, he eats, and he dirties his diapers. Maybe he has worked up to a little cooing. But despite this lack of worldliness, his genome, made up of half his mother's DNA and half his father's, already shows some epigenetic markers from a phenomenon called genomic imprinting.

You inherit two copies of each gene from your respective parents. But in some cases, scientists were surprised to learn, one of those two copies is turned off. Take the gene for insulin growth factor 2 (IGF2), a hormone that plays a big role in gestational growth. Although you inherit a copy of this gene from each parent, only the copy from dear old Dad will be expressed. The maternally inherited allele is silenced. In contrast, cyclin-dependent kinase inhibitor 1C, a gene that is thought to suppress tumor growth, shows the opposite pattern of expression: Dad's copy is turned off, Mom's is expressed. The cases, in which you see these parent-of-origin effects, in an estimated two to four hundred genes, are called "genomic imprinting."

"The phenomenon is paradoxical," explained Catherine Dulac, a Howard Hughes Medical Institute investigator studying genomic imprinting at Harvard University. "It's a huge advantage to have two copies of each gene. But here you have something that shuts down one of the two copies of an essential gene. There must be some advantage."

David Haig, an evolutionary geneticist at Harvard University, hypothesizes that genomic imprinting is simply an evolutionary battle for nutrients, that is, a genetic conflict between the two sexes that helps to determine the size and growth of offspring. Going back to that cute, cuddly newborn—his mom knew he was hers from the get-go. After all, she carried him in her belly for nine months and some change. Dad, however, has no real way of knowing, short of trust and a modern DNA test, that he is the father of that baby. Given this discrepancy of knowledge over the past few million years, Haig suggests, maternally expressed genes are working to make sure all of a mother's various offspring get the resources they need for survival, while allowing the mother to remain healthy and whole enough to have more babies in the future. The paternally expressed genes, however, are not worried about Mom or the health of any other kids that may belong to different fathers. Instead they work to demand more nutrients for that single newborn from the mother in utero and beyond so it might have a leg up on its potentially unrelated siblings.

Take IGF2, the paternally expressed gestational growth gene I mentioned above. If the maternal copy of this gene was not silenced, moms would likely end up birthing some big ol' babies—too big to successfully nourish without detriment to herself as well as to her other children. Haig hypothesizes that while the paternally expressed genes encourage the growth of offspring, the maternally expressed genes help keep growth to a manageable size. He calls this theory the "conflict hypothesis," and some scientific work in animal models lends credence to the idea. "Genes of paternal origin make offspring grow larger and demand more resources from the mother," Haig told me. "But the maternally expressed genes show preference to the mother's ability to reproduce in the future and help limit any one offspring from taking too much."

Why did I lead you down this particular garden path? There has been no evidence that the baby or his parents did anything to change gene expression, and no evidence of life experiences adding methyl groups or throwing kinks into microRNAs with genomic imprinting. But this is a basic epigenetic result that can have profound impact on behavior. What's more, it is an effect that is passed from one generation to the next. Without even taking a bite of cheeseburger, my parents were giving me an epigenome that made alterations to the way my genes expressed themselves and my subsequent behaviors through some of these imprinted genes. Because not only do they affect growth; imprinted genes also have a lot to say about brain development and function. "When scientists have genetically manipulated imprinted genes, the most frequent effect identified relates to embryonic growth," said Dulac. "But the second most frequent phenotype identified involves cognitive function."

In two papers published in an August 2010 issue of *Science* magazine, Dulac, Haig, and their colleagues reported their finding of differential expression of parent-of-origin genes in the mouse brain. They found 347 genes with sex-specific imprinting features that influenced the development of different areas of the cortex and, interestingly enough, that randy-dandy hypothalamus.[2] These findings suggest that imprinted genes are involved in feeding, mating, and social behaviors—like our old friends, sex and love.

Imprinting is not just a simple, static change made in the womb. No, these imprinted genes turn on and off over time, regulating the expression of genes at different points in the life span. What's more, maternally expressed genes contribute most to the developing brain, while paternally expressed genes seem to do more work once the brain reaches adulthood. Why, exactly, is unknown.[3] "[Genomic imprinting] is a dynamic process," said Dulac. "It's not something fixed through the life of the organism. The neurons and neuronal precursors have a certain repertoire during development where maternal or paternal genes are preferentially expressed. Later in life, that pattern of expression is different. It's a major mode of epigenetic regulation and a gold mine for the future understanding of how our genes may control our behaviors."

As I said, it is all my mother's fault. It is possible that various traits and aspects of my behavior, including those involved in love and sex, can be traced back in part to the expression of Mom's imprinted genes. But if I am to give credit where it is due, apparently my father had quite an influence. So I will let him share some of the blame too.

What about Those Cheeseburgers?

I can guess what you're thinking: "Genomic imprinting is interesting and all, but can you really say it's your mother's or father's fault? And what does that have to do with cheeseburgers, for that matter? Or love?" You are right. There I go talking about genes as if they are little General Pattons directing the troops. It's a hard habit to break. But imprinting is important. It shows you how epigenetics can change the way your genome works before you even make the jump from embryo to fetus, not to mention from early childhood to adulthood. But the cheeseburgers may play a role too, if agouti mice have anything to say about it.

Agouti mice are a strain of laboratory mice often used as a model to study diseases like diabetes, obesity, and cancer. As such, it is probably no surprise that they are quite fat and susceptible to disease. They also happen to be a very distinct shade of yellow. (The color is quite reminiscent of first morning urine.) These mice are this way because they have a certain permutation of a gene called the agouti gene. And that particular permutation has proved to be quite resilient. When this strain of mice breed on their own, their offspring are also fat, yellow, and prone to health problems, propagating the same variation on the agouti gene from generation to generation. Jirtle and one of his postdoctoral fellows, Robert Waterland, wondered whether they could change the way the agouti gene expressed itself in these animals without genetic engineering or drug treatment. They opted to try changing a simple environmental factor: diet.[4]

Jirtle and Waterland simply adjusted the feed of a group of agouti mice. Instead of the usual mice chow, they offered a diet rich in methyl donors like folic acid, vitamin B, and choline. As it so happens, you can find these methyl donors naturally in victuals like onions and beets and in prenatal vitamins. Because we already know that methyl groups can bind to DNA and change it in an epigenetic manner, Jirtle and Waterland hoped they would see a change in agouti expression. They kept the mice on this diet and then allowed them to breed.

The resulting pups from methyl-rich-fed parents were smaller, fitter, and a more typical mousy brown. They were also much less vulnerable to conditions like obesity and cancer than their parents. The differences from mom to baby were fairly dramatic, yet the agouti gene hadn't changed. The pups still had exactly the same variation as their fatter, yellower moms. Those methyl groups present in the diet simply attached to the agouti gene and suppressed its expression, allowing the pups to be healthier and brown. "When you see an effect like this, it changes everything," said Jirtle. "You will never think of genetics the same way again."

It certainly brings a whole new meaning to "You are what you eat" (and makes me want to make sure my kid gets his recommended daily intake of methyl donors). This is where the cheeseburgers come in. A lot may come down to my mother's preference for that particular fast food while she was pregnant. Who knows what kind of epigenetic markers that hankering left on my genome while I was still in the womb, changing my body and my health? I am not sure I want to know. And since I doubt that particular fast-food restaurant will be funding any studies to find out, I am not likely to either. The important take-away is that even something like nutrition, in the womb and beyond, has the power to alter gene expression—and consequently brain development and behavior. Cool (and somewhat scary), no?

Early Life Experience

Epigenetics goes beyond imprinting or food.[5] Your early life experience also plays a role. For example, a mother's level of care and affection during the early years has the power to make serious changes to the epigenome. Certainly there is a lot of evidence suggesting that appropriate care in early life is important to behavior. In the 1950s Harry Harlow, a psychologist at the University of Wisconsin, separated newborn monkeys from their mothers and put artificial wire or cloth "surrogates" into the baby monkeys' cages. The researchers soon learned that sustenance was not enough; in order to thrive, those babies needed a soft touch. Without regular cuddling, the babies demonstrated almost autistic-like behavior: rocking themselves, making odd noises, and avoiding novel stimuli.[6] This phenomenon is not unique to monkeys. Children who were the unlucky residents of Romanian orphanages, neglected and left isolated in cribs, during Nicolae Ceausescu's rule also showed a host of mental, physical, and emotional disabilities later in life. At the time these differences were chalked up to strict nurture effects. But the work of a Canadian psychobiologist at McGill University named Michael Meaney suggests that epigenetic effects are at play too. In a landmark study Meaney demonstrated that differences in early maternal care could change the way a certain stress-related gene was expressed in offspring.

Rat mommies do not show their love with expensive toys, extra hugs, or regular trips to the park. They focus instead on a lot of licking and grooming until their pups are weaned. This nurturing behavior does not just keep the baby clean; it also enhances growth and development by facilitating hormone systems. If you threw together a group of rat mommies (often called dams), you would see a wide variety of different licking behaviors. Meaney and his colleagues compared two groups of dams, those that spent a lot of time licking and grooming their pups (high LG) and those that did not (low LG).[7]

When the researchers observed the offspring of these rats, they noticed a few interesting things. First, once female pups grew up and had their own litters, they tended to show the same kind of licking strategy as their mother did. If they were licked a lot as babies, they licked their own babies a lot too; if they did not receive a lot of licking attention when they were younger, they showed the same kind of laissez-faire behavior toward their young'uns. It was a very stable, predictable effect. Grandbaby rats also showed the same licking strategy. What's more, this happened even when rats were cross-fostered with a different type of mom. So even if a rat's biological mom was a big licker, if she was raised by a low licker, she would end up as a low licker herself. Behavior was somehow trumping biology.

Meaney's group also discovered that low-LG pups could not handle stress as well as their high-LG peers. Though all the rats would startle at an unexpected loud noise, the pups that had attentive moms could handle it, quickly going back to whatever rat-type activity they had been up to before the stressor sounded. Pups whose moms did not lick them as much, however, weren't as resilient, cowering in fear for some time after the noise had abated. High-LG pups also showed increased learning over the low-LG pups. Meaney wondered if these differences, with significant transgenerational staying power, could be the product of a gene-environment interaction. He partnered with Moshe Szyf, a scientist studying epigenetic effects in cancer, to find out.

Sure enough, when the group looked at the rat genomes, they found that differences in observed behaviors were linked to the amount of circulating stress hormones, particularly in one type of stress chemical called glucocorticoid. Looking deeper, Meaney and Szyf discovered that a mother's licking and grooming removed methyl groups from the off-spring's glucocorticoid receptor gene. A mother's extra licking attention allowed the production of those receptors to flourish, taking up all that extra glucocorticoid and making for a more mellow, easygoing rat.

Are these results applicable to humans? It is unethical to study humans in the same way we do rats. No one is going to ask moms to neglect their kids so scientists can try to measure their stress hormones in a controlled setting. But given that a history of childhood abuse has been strongly linked to stress and depression, it makes sense that scientists would see the same kind of changes in humans. To test the idea, Meaney's group looked posthumously at the glucocorticoid receptor gene in suicide victims. Looking at DNA from hippocampal cells, they discovered that suicide victims with a history of childhood abuse showed methylation at this gene site—even though they were long into adulthood. Those with happier childhoods did not show the same kind of epigenetic marks.

When I spoke with Szyf, he told me the same thing I learned from Jirtle regarding epigenetics: the genome is the hardware and the epigenome is the software. But he added, "The programmer is the mother. She may not know that she is, but she is." Szyf argues that a mother's behavior is a signal to the child about what kind of environment to expect. "If she gives her child a lot of fatty foods, that is one signal. If she gives him vegetables, that's a different signal. These different signals result in programming, programmed changes in the epigenome to prepare that child for the world around him."

It might be easy to suggest that stress-related programming is permanent—if not "hard-wired," then certainly very stable. This is the case with methylation in genomic imprinting and other types of epigenetic programming. Yet it is not irreversible. When Meaney and his colleagues paired the low-LG rats with high-LG mothers after they reached puberty or put the animals into an enriched environment (a cage with running wheels, toys, and other stimulating artifacts), they saw a disrupted inheritance of the low-LG behavior: the rats behaved more like the high-LG moms when they had their own babies. It would seem there is quite a bit of plasticity, or malleability, in these systems; environment and biology tightly couple to result in learning and behavior. "This is not deterministic either," said Frances Champagne, a former student of Meaney's who now has her own epigenetics laboratory at Columbia University. "There are layers of information around a gene that affect how that gene will be expressed. And these confer plasticity. That's the key."

The brain is always changing. Every experience, interaction, and relationship has the power to change the relative connections between neurons and, by extension, change the circuitry of the brain itself. Your brain at birth is not the same as it is in adolescence, nor the same as it will be in adulthood. In the past decade neuroscientists have been astounded to learn just how plastic the brain is. Previously it was believed that once the brain was done developing—sometime in the teen years—its structure was set in stone. But new research shows that experience has the power to transform the brain at any age, particularly at the molecular level. And those small changes can add up over time. Epigenetic changes are just one way those transformations can occur.

Meaney and his colleagues discovered that high licking and grooming behaviors in mothers show another interesting effect. Not only can they help predict the parenting style of the offspring, but they can also play a role in that baby rat's future mating behaviors.

"High levels of maternal care lead to high levels of maternal care in daughters. It also leads to reduced sexual receptivity. Whereas the offspring who have low levels of care have heightened sexual receptivity," Champagne told me. "So there seems to be a trade-off in different aspects of reproduction as a function of maternal care."

Attentive moms produce pups that will grow up to be not only prolific lickers themselves but also somewhat prudish in demeanor. It is going to take more than just a "How do you do" for a male to mate with this girl. She will take longer to reach sexual maturity, and once she gets there, she will be less receptive to sex. She is picky and takes her time. She will make the boys work for it. In contrast, low-LG moms produce female offspring who are hot to trot. They are not only more receptive to sex, they actually are more likely to solicit it. Nicole Cameron, a student of Meaney's who spearheaded this work, soon discovered that maternal licking and grooming behaviors do not just epigenetically alter stress gene receptor expression; they also make changes to a promoter for a type of estrogen receptor. To the group's surprise, they found that low-LG mothers showed an increase in this promoter, which is probably responsible for these differences in observed sexual behaviors. Again, when animals were cross-fostered, the group saw that the effects were determined by the mother's behavior, not her genes. It was not being the biological offspring of a high-LG or low-LG dam that made the difference; it all came down to experiencing a particular level of licking and grooming in early life.[8]

Since Meaney and Szyf started this line of behavioral epigenetic research, more scientists have gotten on board. I could list a dozen more studies that talk about the different ways epigenetics may affect behavior, not just in early life but across the life span. It is likely that every significant experience and interaction we have with others has the power to change our biology. Some of the most pertinent studies will be discussed in the following chapters. But I introduce the concept here to illustrate how important it is. As much as people want to dispute all manner of individual traits and behaviors in terms of nature *or* nurture, there is no real way to tease the two apart. Our biology impacts the way we are built, the way we perceive what is out there, and the way we interact with the world around us. Our environment provides the sensory information to help the body adapt and respond, ultimately changing the underlying biology. And then the circle goes back around again. How much of nature or nurture has certain effects on our behavior is incredibly context-dependent—and likely impossible to determine in a quantitative fashion. Szyf jokes that it is much easier to explain the concept of epigenetics to your average mom than to a guy who has been studying the finer points of molecular biology for ten years. Moms immediately understand it, especially if they have a few kids who all grow up to be unique individuals despite similar biology and home environment. But it has not been as easy for those invested in genetic determinism to get behind the theory.

"These effects have been completely dismissed by geneticists in the past. Now we realize they are powerful forces that can be modeled both mathematically and experimentally. That changes the whole picture," said Szyf. "You can't just study the cell anymore. There isn't just a cell. The cell acts in a body, that body has a brain, and that body acts in an environment. You can't disconnect them. Our family, our community, our city, our country, our world—all of these are important to understanding biology." And, as it so happens, they are important to understanding behaviors related to sex, love, and parenting too.

I have a confession to make. As much as I may like to jokingly blame my mother and her penchant for greasy burgers for my insatiable love of carbs or my affection for dark-eyed men who will ultimately break my heart, it is only one factor in thousands, millions really, that have shaped me into the person I am, the way I behave—and, by extension, the way I love. There is no separating my biology—my brain—from my environment. And that offers no easy explanations, no general guidelines or rules, when it comes to my relationships.

——————

Chapter 5

Our Primates, Ourselves
(or Why We Are Not Slaves to Our Hormones)

A few months ago a friend showed me an ebook she had downloaded to help explain the innumerable changes her daughter might experience during puberty. It was a modern-day version of the old-fashioned *You're Growing Up!* pamphlets we received in our own preadolescence, often at the doctor's office or in health class, filled with illustrations of pink fallopian tubes and pubic hair growth. As I looked through it, I was struck by the content. In one chapter the author wrote, "Your brain guides all of these changes by using chemical messengers called hormones. Hormones affect many different parts of your body. . . . These chemicals also work on your thoughts and emotions and will affect everything you say and do."[1]

Everything you say and do. From our earliest years we are told that hormones can influence everything, from our boobs (or balls, as the case may be) to our brains to our behaviors. Hormones will flood our system. They will take over and rage out of control. And that impact goes beyond your teenager's extreme behavioral ups and downs. We talk about hormones to explain why a two-year-old boy likes to roughhouse, why our adult moods swing high and low, and why a friend markedly ogles a member of the opposite sex. Our hormones, they say, are a motivating force behind our desire to seek out love and consummate it. But the key word here is *motivate.*

The two hormones that help us along the journey from childhood to adulthood are also implicated in love and sexual behaviors: testosterone and estrogen. The two are often referred to as if they were sex-specific—

testosterone for the boys and estrogen for the girls—but as with oxytocin and vasopressin, it is not quite that simple. Both sexes have ample quantities of each in the body, as well as several receptor types for each across the brain. They work both together and apart to help us grow and mate.

Donatella Marazziti's work demonstrated that there were no significant changes in estrogen or progesterone in people who fell in love. But she did find changes in testosterone levels.[2] It is possible that sex hormones are true to their name and are more involved in mediating sexual behaviors than in the gushy love stuff.

That's right—*mediate*. Our behaviors are not regulated by our hormones. This becomes clear when you see an animal truly in thrall to estrogen and testosterone, like a common rat. Hormones actually control sexual behavior in this species—not mediate or motivate, but control. The female rat ovulation cycle lasts four to five days. When the female is at her most fertile, her hormone levels rise and her back arches up, exposing her private parts to the world. This reflex is called lordosis. It is a sign to all the boy rats that the girl is ready to go (not to mention that lordosis makes it a heck of a lot easier for the male to mount on up to get the deed done). The female rat does not have to consider whether she feels up to sex after a long day of running mazes in the lab. There is no worry about emotional readiness or whether she looks fat. Her hormone levels let her know it is time to get busy. So she does. Otherwise she cannot be bothered. It's just that simple.

Males have it even easier. They try to mate any fertile female they smell. Provided there is no other male around, lordosis means our young buck is not going to have much trouble. All he needs is right there, open and waiting for him. In short, hormones regulate what amounts to a sexual transaction in rats. And by doing so, hormones ensure that rats are prodigious breeders.

You can see similar hormonal effects, though not quite as regulatory, in other species too. Female dogs go into heat. Female baboons' bottoms swell and turn red to let the males know they are fertile and ready. In males high levels of testosterone give certain bird species brighter plumage and prettier singing voices to woo the females. These are outward signs of what is happening inside the body and the brain, making it easy for the animals to know when they will get lucky. There are no games to be played when your body openly betrays your hormonal state.

It is not as straightforward, however, in people. Our sexual behavior over the past hundred thousand years has been emancipated from our hormones. Women acquiesce to nookie not only during the fertile days of their cycle, a phase referred to as estrus; we are often quite willing no matter what time of the month it is. And men do not seem to have any issues mounting us even when we aren't ovulating. Don't get me wrong: estrogen, testosterone, and other hormones are still quite necessary for reproduction. Any couple going through fertility treatments can tell you that. But they are not deciding when and where we are having sex.

"Hormones are not absolute regulators of behavior," said Kim Wallen, a neuroendocrinologist at Emory University's Yerkes National Primate Research Center. "The function of hormones is to shift the balance of behavior in one direction or another. The presence of certain hormones doesn't mean you will exhibit a certain behavior but rather increases the probability that you might."

That increased probability manifests itself in interesting ways. Though you may associate ovulation with bloat and moodiness, hallmark symptoms of premenstrual syndrome, research suggests that it can make a woman feel quite sexy. Kristina Durante, a social psychologist at the University of Minnesota, has discovered that ovulation is linked to women's buying more revealing clothing, showing heightened interest in manly men, and unconsciously heading out to the clubs. Once they get to that nightclub, ovulating women are more receptive to male attention. Nicolas Guéguen, a researcher at Université de Bretagne-Sud in France, found that women at the height of their cycle are more likely to accept a dance invitation from a stranger.[3] If they happen to work at a club, say as exotic dancers, they will pull in higher tips after lap dances when they are most fertile.[4] Men consistently rate ovulating women as having more attractive body scent, facial features, and body symmetry. And several studies have now reported that women with high levels of estradiol, a form of estrogen, are more likely to cheat on committed partners.

Men too are influenced by hormones. Though studies of castrated males have shown that lack of testes (and therefore lack of testosterone) does not always stop the ability to have sex or reach orgasm, more or less testosterone has been linked to changes in aggressive behaviors, risk taking, energy levels, and libido.

In humans it is clear that sex hormones work in subtle ways, so subtle that some evolutionary biologists have argued that human females have a "hidden estrus." But hidden or not (and given the results of some of these studies, it doesn't seem quite so hidden), the sex hormones are still influencing behavior—on both the giving and receiving sides. Scientists believe that hormones may be doing this by direct action on the brain.

Receiving Hormones in the Brain

Sex hormones are not just racing around in your bloodstream without direction. For some time scientists have known that there are receptors for these hormones all over the body. While in the library at the Kinsey Institute for Research in Sex, Gender, and Reproduction, I perused the unpublished works of John Money, a pioneering sex researcher at Johns Hopkins University in the 1950s.[5] In one of his unpublished works about sexuality and gender he lamented that we still did not understand the full reach of hormones in eroticism or sexual behavior. His hope was that the future would bring enlightenment. And it has, somewhat. We now know that the brain is full of receptors for sex steroids, and that these hormones act on brain cells both as a mediator—helping other chemicals do their work in the synapse—and through direct action on their own. They can act as neurotransmitters in their own right, and are both manufactured in and working their signaling magic in the brain.[6] But what exactly are they doing? We haven't figured it all out yet.

"It's outrageously complicated," said Paul Micevych, a molecular biologist who studies estradiol signaling in the brain at the University of California, Los Angeles. "There isn't just one way of signaling by estrogen." The same is true for testosterone. Several receptor types have been identified for both sex hormones; however, there are probably several more that have yet to be discovered—and this is not the only complicating factor.

Like oxytocin and vasopressin, estrogen and testosterone are very similar compounds. *Very* similar. In fact throw a little aromatase, a type of enzyme, on an androgen like testosterone and it will change the chemical structure of the molecule—into estrogen. The male brain has a lot of aromatase, and for this reason all that testosterone in the blood may not reach the brain in the same state. Nor is it involved in only reproductive behaviors.[7] "A lot of the activities that we think of being androgenic are really, at the end, estrogenic," said Micevych. "That is, the androgen testosterone is converted to estradiol by aromatase. Estradiol then binds to a receptor in the neuronal circuits that affect behavior. It's not one over the other, it's both."

This can be found in aggression. Many studies have linked high levels of testosterone to aggressive behaviors. We think of it as a purely androgenic phenomenon. But guess what? If you knock out estrogen receptors in the mouse brain, overall aggressive behavior decreases.

So what else do we know about estrogen signaling in the brain? To date, two estrogen receptors have been definitively identified, estrogen receptor alpha and estrogen receptor beta, though there is a third suspect and probably a few others that have yet to be discovered. It is the alpha version, in the hypothalamus, that appears to be directly involved with reproductive behaviors. But without our knowing all the receptor types, true understanding of the molecule's effects remains elusive. Still, in the two existing receptors, it would seem estrogen helps the brain process information. "Estrogens work in the brain like they are opening a gate," said Micevych. "The same chemical and environmental cues are there whether the estrogen is or not. But when a brain is exposed to estrogen, it opens that gate, so to speak, and allows the information to flow to the right spots to help influence a particular behavior."

So perhaps estrogen makes us more likely to subconsciously pick up social cues. It may enhance the sound of an attractive man's voice or enhance the feeling of a woman's touch. Yet Micevych cautions that there is still a lot to learn at the molecular level before we can even think about translating hormone signaling in the brain to the behavioral level in a mechanistic manner. It's a very tricky thing to study, due to those multiple signaling pathways, those uncharacterized receptor types, and the transformation of androgens to estrogens. And that is all before you add in the cross-talk with other neuroactive messengers. Because, wouldn't you know it, estrogen and our old friend oxytocin interact in areas of the brain like the hypothalamus.[8]

Micevych, a tall man with a striking, angular face and gravelly voice, was happy to discuss these matters further when I caught up with him at a neuroscience conference. "It's not even direct cross-talk—the two somehow activate a third type of receptor, called the metabotropic glutamate receptor, in the hypothalamus, the area of the brain that regulates sexual behavior," he said. "Both the estrogen receptor and the oxytocin receptor compete to activate this third receptor. Both estradiol and oxytocin lead to activation of the metabotropic glutamate receptor, and the response is the same whether estrogen activates that receptor first, before the oxytocin, or vice versa. Interestingly, there is no augmentation of the signaling if they are both applied together. It appears to be an internal check on oxytocin signaling by estradiol and estradiol signaling by oxytocin."

Dr. Money might have hoped for more concrete answers by now about the exact role hormones play in sexual behavior. But it would seem research has uncovered as many new questions as answers. And remember, it is not just biology; the environment plays a role too. Sexual behavior, and thus the activity of these hormones, is context-dependent. Environment matters—and it matters a lot.

"Older studies show that if younger men play a game of football and win, their testosterone levels go up," said Julia Heiman, director of the Kinsey Institute. "It's not so dissimilar to what happens when a new monkey of low status is being introduced to a group. As they gain status in the group, their testosterone levels also rise." This brings up a good point. Since it can be difficult to look at the influence of hormones on human sexual behavior, with or without context, how can researchers learn more about how context and hormones interact to result in different behaviors? As it turns out, there is a fair amount to be learned just by watching monkeys going at it.

Hot Monkey Loving

As I said, studying sexual behavior can be a tricky thing. I know I'm repeating myself, but it's an important point. Human beings are not so keen on inviting investigators into their bedrooms, and university ethics boards frown on people having sex in academic laboratories. Maybe we are intimidated by the idea of lab coats and fervent scribbling, or maybe it's the idea that our sex lives are no one else's business—but despite the fact that sex is openly discussed in many of our favorite magazines and television shows, we are still not all that comfortable talking about it ourselves, let alone examining the specifics. Certainly not in an unbiased scientific fashion, at least. While my girlfriends and I have dissected a few sexual acts over cocktails, I could not tell you with any accuracy how many sexual partners each has had, or which acts, for them individually, constitute "sex." Even as sex becomes a less taboo topic for discussion, we still seem to be holding back. After all, we were raised to believe sex was private, if not outright naughty, and it can be hard to break away from those notions.

To date, much of the data collected about sexual behavior have been conducted by anonymous survey. Though this sort of data can give investigators an idea of the general trends and themes in human sexual behavior, they don't provide many details. And it is the specifics that we would like to know more about.

To that end, researchers have started to look elsewhere, namely, our evolutionary cousins, the primate species. You see, monkeys do not have the same kind of hang-ups we humans have about sex. They seem to have an implicit understanding of sex as a natural act that occurs without judgment and are indifferent to investigators watching them get it on. They don't freak out about what their parents, their friends, or the researchers might say if they knew what they were doing, how they were doing it, or who they were doing it with. And they do not appear to worry about avoiding pregnancy, whether their penis is large enough, or whether those five extra pounds will totally turn off a new partner. They pretty much just do it. Refreshing, no?

Researchers interested in sexuality can observe these animals to find out more about the exhibited behaviors. They can also use animal models to look beyond strict behavior and measure things like associated hormone levels and brain activity. This type of study is by no means a perfect comparison to human sexuality, but until we are more willing to open ourselves up for observation, it will have to do.

To learn more about the ways primates help us understand sexual behavior, particularly sexual motivation, I visited the Yerkes National Primate Research Center to meet with Kim Wallen. I wanted to discover more about what stripping away thoughts of mood, overly sagging scrotums, and general "appropriateness" can help us uncover about human sexual behavior. Wallen, a rugged-looking man with a well-kept salt-and-pepper beard, walked me over to an established group of rhesus macaques to observe what would happen when four new males were introduced to the set. "You're lucky," he told me while climbing up to a widow's walk that would allow us a bird's-eye view of the enclosure. "This is what amounts to excitement around here."

Almost immediately Wallen pointed out a brown-and-white female who had confidently plopped herself down in front of a male in the corner of the enclosure. When one of the human attendants got too close, she used her hands to make threatening slaps and gestures toward the fence. "See her hands? She's actually soliciting the male for sex. She's using those humans as a foil, threatening as if there's a real challenge out there," Wallen explained. "I've never completely understood why it works, but females often use aggressive and threatening gestures to others to trigger the male to mate."

"Maybe to seem more attractive," I suggested, "to play the damsel in distress and make the male feel like he's needed."

The male, a handsome, red-faced gent whom I immediately nicknamed Casanova, did not fall for her ploy. He put his head down and coyly moved it to the side. "That's a groom solicitation," said Wallen. "He's saying 'I know you're interested in sex, but I just really want to be groomed.'"

"I just want to cuddle," I added with a grin. "I'm not a piece of meat."

The female complied, if not impatiently. She started picking at Casanova's head, throwing in the occasional signal that she was still up for a good time once he was ready.

"My impression from watching lots of rhesus engaging in sexual behavior is that females are intensely interested in the sex, and males just want to be groomed," said Wallen. "We tend to think it's the males out there pursuing the females to try and get them to have sex. In fact the females have to work pretty hard to convince the males."

The female stopped grooming and once again showed Casanova her backside. He stared off into the distance, as if he was not getting the rhesus equivalent of a lap dance with a guaranteed happy ending. Once she sat back down, he motioned with his head—another grooming solicitation. I could almost hear the female's resigned sigh as she turned and resumed grooming, albeit a little less enthusiastically. This dance, female butt-showing for sex and male head-nodding for grooming, happened three more times over the next few minutes. New to this group, Casanova had already mastered the art of playing hard to get.

A male rhesus macaque monkey who bears a striking resemblance to my friend Casanova. *Photo by Kim Wallen, Yerkes National Primate Research Center.*

In the rhesus culture the females initiate and control sex. They want what they want when they want it. The males, if they want to have sex, are expected to submit to the females' whims and desires. However, the males aren't completely powerless in this scenario. They can and do say no. Like my friend Casanova. The new kid in town, he watched the landscape carefully. Given that rhesus colonies are a bit of a female club—Casanova would not be an accepted member of the family until welcomed by females of the group—he knew it was in his best interest to get the lay of the land before agreeing to any hanky-panky.

All of a sudden we heard a scream. A second female, screeching and jumping, scared off Casanova's first paramour. Once she was sure the competition was gone, she sat herself down in front of him and made her own play for his affections. Casanova looked confused for a moment, then resumed his posture of disinterest. A slow turn of the head, another grooming solicitation, was the only acknowledgment that Casanova noticed the change of company.

"The notion that the only thing males have on their mind is sex pretty much disappears in this context," Wallen said with a laugh. New to the group and still figuring out the social structure, Casanova had pretty low testosterone levels. Until he found his place within the community, his hormone levels would remain on the low end. But it was not as if the boy had no testosterone coursing through his body, and one would think that, with several sex options available, his hormones would push him to do the deed even if he was the new kid in town.

Another back presentation by the female, a stronger "groom me" request, and these two monkeys had my unwavering attention. I felt as if I were watching an episode of the old sitcom *Friends*. These two monkeys were having their own Ross and Rachel moment right there in the rhesus compound. Getting impatient, I wondered how many times would this female need to let Casanova know she was interested? How could he just ignore that kind of play for his attention? More to the point, what was it going to take for these two to stop playing games and just do it already? I said as much out loud when I saw one of the other newly introduced males already getting busy in another part of the paddock.

"It really is like watching a soap opera sometimes," said Wallen, laughing. "There's a lot at stake here. If he mates with the wrong female and the other females in the group reject him, he's dead. In the wild he would be forced out of the group. But here the group would harass and attack him to the point of injury."

The female finally decided she was not interested in playing games anymore. She ignored Casanova's request for grooming and instead flashed her backside again impatiently. And then again. Then a third time. Casanova simply stared off into the distance, head cocked at a coquettish angle.

"They are in a stand-off now," said Wallen. "She knows what he wants, but she's not going to give it to him." She was ignoring Casanova's repeated "groom me" requests. She was going to wait it out. Just when I thought she had given up, she solicited him again for sex. Another "groom me" request. I groaned. Who would finally give in? We were at an impasse, and neither Casanova nor his sweetheart seemed to be interested in making the first move to overcome it.

Then, just as I thought hope was lost, the female caved in and gave Casanova a swift, cursory grooming around his face. The motions took no more than ten to fifteen seconds. If I had not been watching so intently, I might have missed it. Just as quickly, she presented her back to him again. No response from Casanova. Unfazed by all the rejection, she presented again. Casanova stubbornly responded with another grooming request. This little monkey was determined. He was not going to give in, no matter how tempting, until he was good and ready.

"I think the pattern of promiscuity in rhesus may reflect what the pattern of human sexual behavior would be if you removed cultural constraints," said Wallen. "It's not hard to take promiscuity and shape it into something that looks like the sort of monogamy we humans have ended up with." He paused. "Human monogamy is much less strict than we think it is."

Stubbornly, with obvious irritation, the female presented again. It was as if she were saying, "Okay, buster, last chance. Take me to bed or lose me forever." Instead of asking to be groomed this time, Casanova simply looked away and started grooming himself. It was the ultimate dismissive gesture: If you aren't going to groom me, I do not need you. The female stalked off, probably in search of more amenable company.

Casanova was not alone for long. Before I could even comment about his last girlfriend's abrupt departure, a third female sat down nearby. And this one was a lady; she solicited him very subtly, like a Victorian woman daintily showing her ankles as she took her seat. Casanova again solicited grooming. She acquiesced, but only for a moment. It was easy to see that grooming was not her thing. Lady or no, she was looking for some action. Once again Casanova appeared completely uninterested. I sighed with vexation.

"What do you think of this?" Wallen asked.

"Honestly, I'm wondering what the heck Casanova is waiting for!" Three separate females, all good-looking enough in their rhesus way, had offered him what every male supposedly wants: sex. How dare he turn them all down?

"This is fairly typical," Wallen explained. "It may take a male and female an hour and a half before they actually do it. Even then, they may start mating, stop, break apart, and then come back together. It may take an hour or more after mating starts before he actually ejaculates."

The thought exhausted me. With all these games, even in the rhesus population, how do babies ever get made? But before I could start yelling at Casanova to stop posturing and just hit that already, shrieks erupted below me. Jumping and screeching with menace, a large group of rhesus barreled toward the corner where Casanova and his group of lady friends had been hanging out. They were upset with a different animal, one of the other new males, and massed in force to teach him a lesson. They did not seem to mind if Casanova ended up as collateral damage. Wallen ran down to break up the fight before one of the animals was injured.

When calm was restored, it took me a minute to find Casanova. I soon spotted him hiding under a large climbing structure, scared and alone. No females were approaching now. Once Wallen returned to the widow's walk, he explained that Casanova abstained from sex for good reason: with the introduction of four males to the group at one time, there were a lot of uncertainties about what the new members would do to the existing social structure. "Now we know why he was so hesitant," said Wallen. "He was scared out of his wits. It's really hard to get an erection when you're petrified that you're going to get attacked."

The highest-ranking member of any rhesus group is going to be a male. But it is still a girls' club with girls' rules. It is the alpha female who picks the alpha male by mating with him. If he offends her in some way, he can always be replaced. While our eyes were on Casanova, the alpha female was still trying to decide who was worthy of her affections. The monkey under attack was the male who made the faux pas of getting a little something before she had made her choice. The social structure of the group, including its rules, culture, and leadership, had yet to be decided. It was in Casanova's best interest to wait it out—not only in case the alpha female decided to bestow the ultimate prize on him, but also to make sure he was not inadvertently offending someone who had pull once order had been established. Sex is not the ultimate prize when it can get you thrown out of the group. Despite all those hormones running amok in my friend Casanova, he was able to keep his head and think about the consequences. Even a monkey has the power to overcome his hormones.

Society's Pull on Our Hormones

Now that the Casanova Show was over, Wallen and I made our way to a different area of the field station, where one of his graduate students, Shannon Stephens, was observing a smaller group of rhesus monkeys. In this tighter enclosure the animals could run and frolic, seemingly oblivious to us as we watched from above. To my untrained eye these monkeys looked no different from the ones Wallen and I had watched earlier. But a few of them, with dyed patterns on their backs, had brains that were significantly unlike the others'.

Shannon was studying behavioral differences in these animals after neonatal amygdalectomy, a surgical procedure that extracted their amygdala after birth. In particular she and Wallen were interested in whether the removal of this almond-shaped brain region responsible for emotional memory would affect the onset of puberty in young females.

The average age of human menarche, or the start of menstruation, has decreased in the past hundred years. Some epidemiological studies have suggested that the abundance of fatty foods in our diet or the increased amounts of hormones we ingest may be at play. But just as much research has focused on the role of social environment. Sexual abuse, the presence of a nonrelated male figure such as a stepfather living in the home, and exposure to sexual stimuli have all been linked to early menarche. Later menstruation seems to happen in larger families or when a girl has a close relationship with her biological father, though the why's and how's of this phenomenon are unknown. There's that context again, coming in and messing with our hormones. Taken together, the findings suggest that our environment can alter the way our hormones are expressed—and, by extension, our behaviors.

Though there is no specific hormonal event that triggers a girl's first period, a rising level of estrogen has been noted. It is the same in the rhesus. The flood of estrogen that precedes menarche is a critical step toward the animal's fertility. Social environment too is important. Social rank in the group is a factor that can predict how early the onset of puberty might be. Once again, environment and context are influencing our hormones, not the other way around. "High-ranking females are more likely to go through puberty earlier than the low-ranking ones," said Wallen. "One view of the amygdala is that it stamps social context into emotional memory. . . . So it's possible that since puberty in females is socially mediated, in this case by rank, we questioned what the effect would be if you removed the amygdala."

What happens to hormonal expression when you remove the ability to read social context? In the eight monkeys who received neonatal amygdalectomies, six had already gone through puberty, a full year earlier than females who have an earlier menarche. But more interesting to Stephens and Wallen was that the first monkey to undergo the procedure displayed normal ovulatory cycles but a complete and utter disinterest in soliciting sex, an important social behavior in rhesus females. "She didn't show any interest in the males," said Wallen. "The males would solicit her, but that was it."

Unlike those brazen hussies who audaciously flashed their backsides to let Casanova know they were good to go, Opie, the first amygdalectomized monkey, never picked up the steps to the mating dance. It was unclear if she ever understood enough of the social context to realize there was a dance at all. She had normal cycles and the right hormone levels to have offspring, yet she never approached or solicited males for sex. The lack of amygdala meant that comprehending the social environment, especially when it came to mating rituals, was beyond her. It was a staggering revelation.

Stephen's study is ongoing; the group is waiting to see if the other amygdalectomized monkeys will behave like Opie. But Wallen said that the preliminary results reinforced what he had been saying all along: we are not slaves to our hormones. Many factors, including social context, play a huge role. The hormones talk to us, definitely. But it's not a "must-follow" kind of message. "My view of hormones is that they are really just suggestions," Wallen told me. "From an evolutionary standpoint, all they've got to do is increase the probability that you'll want to have sex at the right time for reproduction."

This rings true with Micevych's statement that, at the cellular level, hormones open the gates so we have more access to information. But they do not insist on a particular course of action, nor do they take away our ability to choose how we behave. Ultimately, no matter what our hormone levels may be or how motivated we are to get busy, we still have the power to decide whether or not we will have sex, even when, as with Casanova, a sure thing is staring us straight in the face. As Wallen said, "We don't have hormonal regulatory mechanisms. What humans have are simply motivational mechanisms. And motivations can be easily ignored or paid attention to, depending on context and environment."

Hormones do a lot of influencing, that's for sure, but we are far from their slaves.

Chapter 6

His and Her Brains

Bill Cosby, comedian and father figure to the Generation X crowd, once joked, "Men and women belong to different species, and communications between them is still in its infancy." It's a great line because it feels like a basic axiom of intergender relations (and would appear so even if Dr. Huxtable had not been the one to utter it). Let's face it, males and females seem to be fundamentally different.

How often have you responded to or dismissed some baffling behavior of the opposite sex with "He is *such* a guy" or "Must be a chick thing"? Even my five-year-old seems to understand that saying, "Because she's a girl," or "Because he's a boy," is an explanation unto itself. This is not misogyny on his part—he does not see it as a criticism—it is just simple rationalization. From an early age we all appear to implicitly understand that boys and girls, men and women, are different. Period. Some experts might even have you believe that men and women are so different that they are destined to always stand in opposition when it comes to both communication and behavior.

Since John Gray published his *New York Times* best-selling book *Men Are from Mars, Women Are from Venus,* we have tried to explain away all kinds of relationship troubles by lumping behaviors or traits into certain categories. Why don't men call when they say they will? It's a guy thing. Why do women lose interest in sex as they age? It must have something to do with their gender. Why do so many relationships fail? Men and women simply cannot relate without some kind of primer on translating each other's behavior. Many have argued that perhaps the biggest problem with sex and love (the heterosexual variety, anyway) is that men and women are too dissimilar to properly connect. Could the problems we face in love and sex come down to the differences between our genders?

In recent years countless scientists and authors have written thousands of pages discussing the distinctions between the brains of the two sexes—and by extension, the sex differences that are often seen in behavior. Some argue that although boys and girls may show natural variations in brain development and structure, nurture can always trump them.[1] Others have promoted the idea of "neurosexism," the suspicion that neuroscientific studies of sex differences in the brain, particularly neuroimaging studies, are simply a modern way to support the old notion that men are somehow intellectually superior to women.[2] We can leave both these concepts for others to support or refute. Instead I will explain what neuroscience has offered in terms of how these intergender brain differences develop and how they may apply to the study of sexual behavior and love.

"Eve Plus Androgen Equals Adam"

John Money was a leading sex researcher at Johns Hopkins University in the second half of the twentieth century. Much of his work involved the study of hermaphrodites, individuals born with both male and female or ambiguous genitals. Extending from that work, he also had quite a bit to say about gender's influence on behavior and how observed differences between boys and girls may develop. In an unpublished manuscript originally intended for *Scientific American* titled "Pygmalion Updated," he succinctly explained how hormones designate gender in the womb: "Eve plus androgen equals Adam."[3]

Catchy, no? But it does capture a rather complicated process in a simple way. Embryos pretty much all start out as female. They just don't always stay that way. If the fetus in question gets a copy of the Y chromosome from dear old Dad, a flood of testosterone, an androgen, rocks the womb early in pregnancy, somewhere between six and twelve weeks. This increased level of testosterone guides the development of the penis, scrotum, and testicles. Without it, the fetus would remain female and develop the corresponding reproductive tool kit.

But androgens do more than just determine what you are packing in your pants. Testosterone, estrogen, and progesterone are also critical to the brain's development. They work in concert with a variety of other proteins and chemicals to help organize the brain into distinct regions and circuits. They are also responsible for sexual dimorphism, or systematic differences in form, in several areas of the brain.

The Sexually Dimorphic Brain

Historically when people talked about sex differences in the brain, the focus was on reproduction. It probably comes as no surprise that the hypothalamus, that little relay station to the pituitary gland implicated in all manner of sexual behaviors, is different in males and females. In rats a small cluster of cells in the hypothalamus called the sexually dimorphic nucleus has long been known to be significantly larger in male rats than in their female counterparts. What do I mean by "larger"? I mean that the actual volume of this area is larger relative to the overall volume of the brain it resides in. Work in animal models suggests that the bigger a brain area is, the more important it is. For example, rats rely more on their sense of smell than their eyesight to navigate the world; thus their olfactory brain areas are proportionately larger than their visual ones. Humans, who depend more on sight to get around, show the opposite effect when it comes to the relative sizes of these brain regions.

The human analogue to the rat's sexually dimorphic nucleus is called the third interstitial nucleus of the anterior hypothalamus. This mouthful of a brain area has been linked to male sexual appetite as well as overall sexual behavior. Like the rat's, it is bigger in men than in women. Given the variations observed in the outdoor plumbing, it is no surprise that you would see some sex differences in the brain regions associated with them. And for decades most researchers believed that was all there was to it.

New research, however, demonstrates that there are sex differences across the entire brain, and the brain areas affected by these gender-specific, in-utero hormone baths influence a heck of a lot more than just reproduction. Brain regions involved in emotion, memory, learning, perception, executive function, and stress response also show some differences between the sexes. Jill Goldstein, a neuroscientist at Harvard Medical School, used neuroimaging to look at sex differences across the whole brain. She and her colleagues discovered that parts of the frontal cortex are bigger in women than in men, as well as several brain regions in the limbic cortex, the emotional response area of the brain. In men, the parietal cortex and amygdala, on average, are larger. These areas are involved in spatial perception and navigation and emotional arousal and salience, respectively. Even more interesting is that all of these areas show a very large number of sex hormone receptors during embryonic development. Those hormones in the womb are making some important changes. And when puberty brings another hormonal onslaught later in life, these circuits are activated and ready for business.[4]

As I said, the long-held assumption is that the bigger the area, the more important it is to the organism in question. As researchers looked deeper, they found that these larger areas also demonstrated greater densities of neurons as well as more dendritic growth, which is an indicator of neural reach; the more dendrites on a cell, the more synapses it can form. Did these variations designate a difference in cognitive processing between the two sexes? Anecdotal evidence certainly suggested so. But it was not until neuroscientists started using fMRI during cognitive tasks that they knew for certain.

Larry Cahill is a researcher at the University of California, Irvine, who studies emotional memory. He often uses neuroimaging studies in his research. Several years ago he noticed an interesting quirk between male and female study participants during an fMRI study. He already knew that the amygdala was usually larger in men and therefore suspected that it might show more activation during certain kinds of emotional tasks. But he was surprised to find that men and women showed varying baseline activation in the amygdala. That is, when the study participants were only relaxing inside the magnet (as much as one can relax inside an fMRI), their amygdalas were lighting up in different ways.

This was a big deal for cognition and for the study of love-related behaviors. Many of the neuroimaging studies found in the scientific literature, including some that look at love and sexual behavior, take blood flow measurements in only one gender. Any conclusions drawn from the results have to be reconsidered when you realize that male and female brains are just a little bit dissimilar.[5] In light of Cahill's work, it would be all too easy to throw the baby out with the bathwater and turn your nose up at any neuroimaging study that scans only one gender. Not to mention that the idea of "his" and "her" brains is quite the political topic. In this postfeminist world, all too often the notion of "different," as demonstrated in scientific study, morphs into "better" or "smarter." But researchers like Cahill and Goldstein are careful to point out that just because male and female brains might develop from a slightly different blueprint, their behavioral output may not differ all that much. Take stress, for example. "Male and female brains work slightly differently under stress, even though this may lead to similar levels of performance," said Goldstein. "I often use the example of a Mac and a PC computer. They have been built differently, but they can accomplish the same kinds of tasks at a similar performance level."

Goldstein and her colleagues compared the activation of stress response circuitry (the amygdala, hypothalamus, hippocampus, brain stem, orbitofrontal cortex, medial prefrontal cortex, and anterior cingulate gyrus) in men and women, both in the women's fertile phase of their menstrual cycle and in the middle of the cycle, as they viewed a series of very unpleasant photos from a standardized photo set—think gory body parts after car crashes and the like. In women, the study showed different patterns of activation in this circuit between the two phases of their menstrual cycle. Women and men also showed distinctly different patterns of activation. But Goldstein says one of the most interesting things about this study was the fact that all participants reported having the same kinds of stressful feelings when they saw the photos.[6] Different brain activation, but similar feelings. How about that?

"We showed that hormonal status regulates the stress response in the brain differently at different points in the menstrual cycle in women," Goldstein said. "Further, these hormonal differences explained sex differences in the brain's response to stress, even in the face of no sex differences in the subjective feeling of stress. This suggests that hormones are involved in maintaining homeostasis in the brain in response to stress." Basically men and women experience stress in the same way, but their stress response involves different mechanisms in the brain.

I can't help but notice that many of the brain areas in this stress response circuitry overlap with those activated by romantic love. But to date no one is looking specifically at sex differences in love, perhaps because the study of love is so new to neuroscience. But when I asked Goldstein about whether we might see gender differences in love and sexual relationships in the future, she reminded me of how important it is to be conservative when interpreting results from these kinds of studies.

"We know there are numerous sex differences in how the brain develops through childhood and then functions into adulthood," she said. "But how these differences may or may not relate to complicated concepts like 'love' or 'desire' is unknown." She paused for a moment before continuing. "The study of something like love is very complicated and cannot be reduced simply to the neuronal level. It can be thought of on many different levels, both conscious and unconscious, involving everything from emotion to cognition to physiology, psychology, sociology, and 'chemistry,' just to name a few."

When I spoke to Helen Fisher about her neuroimaging studies, she said it would be interesting to take a closer look at gender differences in future neuroimaging studies of romantic love. "People are quick to assume that men and women are very different in this regard, that men avoid commitment and women really want commitment. Certainly, from an evolutionary perspective, it's equally beneficial for both sexes to have a committed partner to help raise offspring. But does the brain back that up? It is something we still need to look at."

Let's Talk about Sex

So the jury is still out on whether men and women approach love differently from a neurobiological perspective. But sex seems to be a no-brainer, pardon the pun. Common wisdom tells us that men and women are very different when it comes to sex. Men are more visual, women are more emotional. Men are willing to have sex with just about anything that happens across their path; women are more selective when choosing sexual partners. And it is often thought that men pretty much want sex constantly, whereas women are more the sexual camel type, able to go without intercourse for quite some time without a problem. This is what we often hear, anyway. If these particular stereotypes hold true, you'd think there'd be some neurobiological evidence to back them up.

Compared with other visual images, sexual images seem to produce a special effect on brain activation, in both men and women. "When we put people in the magnet and show them sexual stimuli, the response in the brain is two to three times stronger than any other kind of image or stimulus I've ever used," said Thomas James, a neuroscientist at Indiana University who works with researchers at the Kinsey Institute studying sexual decision-making processes in the brain. "Sexual photos are incredibly arousing images, just in the general sense of arousal. They really get the brain going." Even if those images do not result in direct sexual arousal, as indicated by noticeable erection or vaginal lubrication, brain activation still goes wild. Our brains, apparently, are fine-tuned for porn.

Are there different reactions between the sexes when viewing those sexual images? Apparently so. Kim Wallen, my favorite rhesus-watching companion, and his colleagues at Emory University noted that men seem more responsive than women to visual stimuli of an arousing nature. When they put both men and women in the fMRI and showed them a variety of kinky pictures, both had similar activation patterns in the reward circuitry. But the left amygdala, an area of the brain responsible for attaching context and meaning to the outside environment, showed significantly more blood flow in men than in women. Limbic regions, correlated with emotional responses, and the hypothalamus, that seat of sexuality, also showed greater activation in the men. Out of all those variations in cerebral blood flow, Wallen and his colleagues homed in on the difference observed in the amygdala and argue that this area may be responsible for the fact that men are more affected by visuals when it comes to sexual behavior.[7]

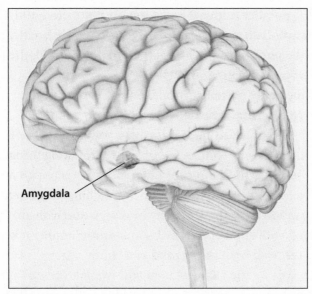

The amygdala shows greater activation in males than in females when viewing sexual stimuli. *Illustration by Dorling Kindersley.*

"What was really interesting to me was that the women in this study actually subjectively rated the pictures as more sexually arousing than the men did," said Wallen. "Yet we still saw higher activation in the men in the amygdala and hypothalamus. That says something."

Heather Rupp, a former graduate student of Wallen's who went on to the Kinsey Institute after leaving his lab, argues that the general neural circuitry that underlies sexual arousal is probably pretty similar in men and women. But, she maintains, those circuits may be differentially activated depending on the kind of stimulus presented.[8]

That is one explanation. A good one too. Of course, there is an alternative hypothesis. Just because people are looking at duplicate photos does not mean they are paying attention to the same elements. Consider a good piece of porn, film or photo; there is usually quite a bit going on there. What if the dissimilarities observed in brain activation were due to the fact that people were just attending to different information?

Rupp and Wallen had fifteen men, fifteen women on the birth control pill, and fifteen women not using hormonal birth control view hundreds of sexual photos from free porn websites. As each photo was displayed, participants were asked to rate the sexual attractiveness of the photo with a number rating, 0 being the least attractive and 4 being the most attractive. If they found the image completely unattractive, they could give the photo a rating of -1; when that happened, the photo was not included in the analysis.

While the study participants made their sexual attractiveness decisions, the researchers recorded not only how long each participant viewed each photo but, using eye-tracking software, where exactly they were looking. Rupp and Wallen found a few interesting things. First and foremost, there was no significant difference between men and women in their subjective ratings of the stimuli or how long they looked at them. They were even as far as these two measures go, which disproves the idea that women do not appreciate visual sexual stimuli; women kept up with the men in terms of both ratings and view time.

Second, despite the fact that men and women showed appreciation for the porn, the two groups were not looking at the same things in each photo. Women tended to rate as more attractive those photos in which the female actors were looking away from the camera. But men did not seem to care which way the female actors were looking. Neither gender bothered to look too long at genital close-ups, but it was only men and women on oral contraceptives who rated them as significantly less attractive. Thus there were differences—sex-specific preferences—even when, across the group of stimuli, men and women chose similar ratings on the attractiveness of photos.[9] Once again, context matters.

Cognitive and behavioral studies have suggested several other key differences between the sexes. When researchers at the Sexual Psychophysiology Laboratory of the University of Texas at Austin looked at how well individuals remembered sexually relevant information in an erotic story, they found gender differences. Sex differences in cognitive tasks are often assumed to correlate with differences in neural activation. Previous studies had suggested that men were more likely to recognize specific erotic sentences—more quickly too—than women. But women were more accurate at recognizing romantic sentences that appeared in the text. In recall studies, men often erroneously remembered things from the story, of both a sexual and a romantic nature. To suss out what was going on, Cindy Meston, head of the Sexual Psychophysiology Laboratory, and her colleagues had seventy-seven undergraduates read a sexual story and then perform a memory task. In this study men were more likely to remember the erotic elements of the story, and women were more likely to recall the specific characters as well as the love and emotional bonding bits. Here the participants played straight to stereotype.[10]

When I asked Meston if she believed there might be a variation of neural activation underlying these differences, as many researchers assume when they see a disparity between the sexes in cognitive tasks, she paused before responding. "I do not know," she said. "One could make the argument that erotic cues are more rewarding for men than [for] women, that perhaps a man gets a bigger dopamine burst when he sees an erotic picture or reads an erotic story. But I really do not know what we would see in terms of brain activation."

Sexual Motivations

Men and women show different amygdala and hypothalamic activation. They don't look at the same things in sexual photos. They remember contrasting details from erotic stories. What about sexual motivations? Do men and women have separate reasons for having sex?

Meston's lab recently examined that question. In a large sample questionnaire study, she and her colleagues looked at all the reasons folks from eighteen to seventy might have for getting busy. The results surprised them. "You always hear that women are more likely to have sex for love, men for physical gratification. And we did see some of that," said Meston. "For example, men were more likely to engage in opportunistic sex and women in sympathy sex. But across that age range, we found many more gender similarities than differences. The top three reasons for having sex were the same in both genders—they were having it for love, for commitment, and for physical gratification."

You heard it here. Sure, gender differences are seen in a variety of studies. Many of them support the ideas we have about the ways men and women view sex. But there are a lot of similarities there too. Subjective reports of arousal, our reasons for having sex, show a lot of overlap between the genders.

"Some of these differences may be explained simply by differences in anatomy," said Meston. "Anatomically, men get an erection when they are aroused. That is a pretty hard thing to ignore. It is a strong, apparent signal grabbing his attention, probably distracting him from other things that he may need to get done. With women, the sexual response is tucked away, and the vagina does not hold as much blood as the penis. It may not be as strong a signal. So in this case, it may be what is going on in the rest of the world that is the distraction, not the arousal itself. Those anatomical differences might explain a lot of the gender differences you hear about."

Love Remains the Same

What about love itself? Is what I experience when I feel love qualitatively different from what a man experiences? If I consider Semir Zeki's hypothesis that literature and art across the ages show a common substrate for love in the mind, I might suggest that descriptions of sex by male and female authors and artists are sometimes different. But descriptions of love by writers of both genders? They aren't all that dissimilar.

Although previous neuroimaging studies of romantic love by Zeki and Fisher included members of both sexes, a precise comparison of brain activation between the two was not undertaken. Zeki and his collaborator John Paul Romaya decided to take a closer look to determine whether there were gender differences in the way men and women experience love.[11]

They compared cerebral blood flow in twenty-four people in committed relationships who claimed to be passionately in love (and scored high enough on a passionate love questionnaire to back that claim). Twelve of those participants were men, and six of those men were gay. The remaining group of twelve women was also made up equally of gay and straight women. The study paradigm was identical to Zeki's initial romantic love study: each participant's brain was scanned as he or she passively viewed photos of his or her partner and a familiar acquaintance matched in gender and age to their true love.

Zeki and Romaya found similar patterns of brain activation and deactivation across all participants, replicating the findings from Zeki's original romantic love study. Once again measurements of cerebral blood flow support the idea that love is both rewarding and blind. But there were no significant differences between activation patterns in men and women. Considering the sexual dimorphism seen in many parts of the brain, it's an intriguing result. It appears that love is love, no matter what gender you are.

When I asked Zeki if he was surprised by the finding, he chuckled. "To be honest, I was entirely agnostic," he said. "I cannot say I was surprised by the results. But I think this is one of these studies where people would have said, 'I'm not surprised,' even if the results had gone the other way."

So Are Men and Women Different or Not?

It is easy to fall back on old stereotypes, to simply say that men and women are poles apart. And perhaps those differences are enough to fuel those storms you commonly see in relationships. It would almost be easier if we could say that male and female brains are just too dissimilar, that they perceive and process love and sexual stimuli separately; it would give us something to hold on to when no other explanation for our love-related woes seems available. Alas, it is not quite so simple.

"When we talk about sex differences in the brain, people want to go all 'Mars, Venus' on you. They want to take these results and try to spread males and females way apart on function and ability," said Cahill. "It is not like that. When you are talking about sex influences on brain function, you may have two bell curves that are significantly different from one another in certain instances. But those bell curves are still overlapping."

Goldstein concurred. "There is more variability within a given sex than between sexes in cognitive behavior and the brain. That is important. In fact, I always say it twice so that people really understand that," she said. "There is more variability observed between women than between women and men in both the size of different brain regions as well as the function."

Meston saw the same kinds of overlapping bell curves in her research. "Every person brings their own individual history to any sexual situation," she said. "The reasons why they are having sex, the way they feel about the sex, and the consequences of having sex are all very different across individuals no matter what gender they happen to be."

That's something to consider the next time you want to chalk up your partner's quirks and shortcomings to his or her gender alone.

Chapter 7

The Neurobiology of Attraction

What is it that attracts us to another person? When I asked my friends what initially drew them to their significant other, the responses varied from "his gorgeous blue eyes," to "her honesty and intelligence" to "He had air-conditioning—we met in Niger in summer." I also heard quite a few comments about beautiful smiles and nice butts, from both sides of the gender divide, as well as tributes to cute dogs, cable sports packages, passion for a career, motorcycles, crack handyman skills, attractive friends, and even sweet, sweet pity. The responses truly ran the gamut. When I consider my own experience, I can honestly say I've been attracted by many things. I dated one man who had a laugh that immediately put me at ease, another whose brilliance placed him at the top of his field. And I see no reason not to mention that I've gone out with a guy or two for no other reason than they were drop-dead gorgeous.

Although many of us can point out the one thing that initially sparked our interest, it is never quite that straightforward. Simply stated, there is no *one* thing. No matter how nice a woman's derriere looks or how hot a Niger summer gets, any attraction is made up of a variety of elements, physical, mental, and emotional—perhaps even physiological. Other types of comments made by friends mentioned butterflies in the stomach, a feeling of being drawn to the person as if by "tractor beam," and "just knowing" this person was meant for them. A couple of friends even admitted that the basis of their attraction was a mystery: they weren't certain what exactly drew them to the person, only that it was undeniable at the time. This is a case where the whole is much, much more than the sum of its parts.

"Obviously, there's a lot to attractiveness," said Thomas James, a neuroscientist at Indiana University who studies sexual decision making. "There's his face, and that's important. But you want to see your guy get up there and dance, see how he moves. You want to hear his voice. He may have a great face, but if it's paired with a squeaky high voice, there goes the attraction. If he has an okay voice, what he actually has to say plays an enormous part. There's a lot to it."

"You can't forget smell," I added. "Smell is important too." I've been overpowered by cheap cologne far too many times not to mention it.

"Right. So we are beyond the visual here. We have the auditory, the olfactory, how the guy moves, what he says—and we're only getting started," said James. "There's so much there. It's a real challenge to try to capture those variables in an experimental way."

And the challenge only seems more immense when you try to examine it from a neurobiological perspective. What is it that we are picking up from another person that forces our attention so strongly upon him or her? That can sexually arouse us after only a look or a word? Make us crave that person's future company? After attraction has been established, what is it that determines whether it will grow into love? Does that attraction have to be there immediately, or can it grow over time? It's hard to know which question to address first.

Here's a case where animal models are not much help. Female rats don't care if a male rat has a sense of humor or what he does for a living. It doesn't matter to male rats how many baby daddies a female has previously entertained or whether she has a season pass to Penn State games. I don't even know how to begin to qualify what might count as hot on the rat ass scale, but I do know that a show of teeth in these critters usually precedes an attack. Courtship in rodents and human beings is not all that analogous.

For example, the prairie vole, the rodent mascot of love, becomes attracted to another vole after a lot of urine sniffing. The pheromones, small chemosensory compounds, in the urine give these animals enough information to help make a connection and get to work getting busy. The same setup is not going to work with humans. Though I made sure to mention smell when I talked about attraction with Thomas James, I've never taken it so far as to smell a guy's pee.

That's not to say that different scents don't have some influence in the human brain. Wen Zhou and Denise Chen, researchers at Rice University, found that human sweat during sexual arousal selectively activates different areas of the brain. The duo had male sweat donors abstain from using deodorant, antiperspirant, and scented body care products before coming to the lab to watch twenty minutes of heterosexual porn and then a neutral video of the same length. The men did their watching with absorbent pads under each armpit that collected the sweat associated with each viewing condition. The researchers then pooled the sweat from all the participants after they watched the sexy video and did the same for the neutral viewing condition to create the study's two olfactory stimuli (that's science talk for a sample of something participants would smell during the study).

Using fMRI Zhou and Chen scanned nineteen women who had sniffed the combined products of the porn and neutral video viewing sessions. They also measured cerebral blood flow after the participants smelled a putative sex hormone called androstadienone, a metabolite of testosterone that has been linked to elevated mood and cognition in women, and a nonsocial control smell. The study participants inhaled each compound for twelve seconds, then rated the pleasantness of each smell on a scale of 1 to 5. Note that these women were not told what the smells were, or even that they were human in origin. They were simply told to take a good whiff and determine how much they liked it.

The researchers found that the human sexual arousal sweat showed a distinct activation pattern in the right orbitofrontal cortex, an area related to socioemotional regulation and behavior, and in the right fusiform region, an area usually implicated specifically in human face and body perception. The group also saw activation in the right hypothalamus, the seat of sexual behavior, from the sexual arousal sweat but none from the control smell.[1]

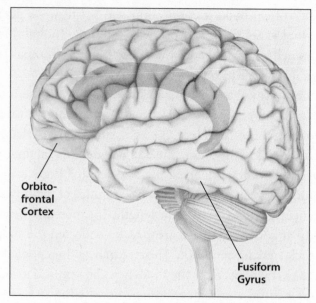

Zhou and Chen found activation in the orbitofrontal cortex and fusiform gyrus when study participants smelled human arousal sweat. The hypothalamus was also activated. *Illustration by Dorling Kindersley.*

This means our brain is doing a lot with the chemosensory information in the sweat of sexual arousal. Somehow we know that it belongs to another human without being explicitly told so. Our brain also seems to understand that the scent has something to do with sex. These smells are managing to convey a lot of important information without our even being aware of it.

Given that we humans do not have a solid understanding of all that goes into attraction, and the assumption that evolution wouldn't throw away a system that works for the majority of the animal kingdom, some neuroscientists have suggested that chemosensory cues play a role in our attractions. Perhaps along with (or even instead of) bright smiles and intelligent conversation, a person's smell lets us know if he or she is worth spending time with. Certainly more than a few fragrance companies and Internet outfits believe this—human "pheromones" have long been advertised as a sure thing for attracting members of the opposite sex. Though some contend there's no such thing as a human pheromone, Zhou and Chen's study suggests that our brains are picking up something from social smells.

What Is a Pheromone?

Nearly 150 years ago Charles Darwin pointed out that being smelly—you know, in the right way—could help male ducks, elephants, and goats procure a mate. It was not all pretty feathers or a strong trumpeting cry. Rather, in *The Descent of Man, and Selection in Relation to Sex* Darwin argued that chemical signals are just as important as visual or auditory ones when it comes to animal attraction.[2] As the old Cole Porter ditty goes: "Birds do it, bees do it, even educated fleas do it"—except that rather than falling in love, they release a variety of chemical signals through their skin and waste products that cue members of the opposite sex to come calling. Those chemical signals would be the pheromones.

Originally defined by Peter Karlson and Martin Lüscher in 1959, a pheromone (from the Greek *pherein,* "to transfer," and *hormōn,* "to excite") is a small chemical molecule that helps an animal communicate with other animals within its species. I know that seems vague. You may be asking, "Communicate what, exactly?" The answer is "many things." Though the word *pheromone* is often used as a synonym for a sexual attractant, that's a bit of misnomer. There are several types of pheromones that have been identified in the animal kingdom: primers, releasers, modulators, and signalers.

Primer pheromones are just that: primers that act on the body's neuroendocrine system and jump-start different functions, such as menarche and puberty. Releasers provide the sexual attractant variety of pheromones, eliciting a behavioral response such as lordosis in females, the arched-back ready position for sex. They can also be alarm signals, motivating certain animal species to withdraw from dangerous situations or gear up for a throw-down. Modulators can change emotions; they are chemicals released by the body that can positively affect mood and emotional state. And signalers are compounds that provide information about gender, reproductive status, and age. Our bodies, it seems, don't keep many secrets. They are releasing little chemical bits of information at any given moment for others to unconsciously pick up. Those little bits may then alter the internal body chemistry or help an individual intuitively size up social situations.[3]

In the fifty-odd years since Karlson and Lüscher first came up with the term, there has been some debate about whether there is a human pheromone. There are many reasons for this, and most are beyond the scope of this book. Some say pheromones work through the vomero-nasal organ, a specialized olfactory organ that interfaces with the brain in lower mammals. But humans don't have this organ; ergo, some say, we don't process pheromones. Some say human olfactory processing has evolved beyond the simple molecule. Like our big frontal lobes, human chemosensory signals are more complex than simple phero-mone molecules. But perhaps the most important criticism of human pheromones is that, as of yet, a unique human pheromone has not been identified. The only candidate, discovered by Martha McClintock at the University of Chicago, is a chemical that helps to sync the menstrual cycles of cohabitating females. Though studies like Zhou and Chen's have demonstrated that humans are susceptible to chemical messages from smells, it's unclear if those chemicals are pheromones in the strict sense of the word.[4]

One chemosensory compound that has been the subject of intense study is the major histocompatibility (MHC) complex. As the name says, it's a complex: a mixture of hundreds of different compounds, perhaps more. Although the MHC complex is often talked about as if it were a pheromone, it doesn't fully meet the criteria. It is a group of genes that leads to a specific odor-print. As it turns out, every human has a signa-ture odor—like a fingerprint. In humans the MHC complex is based on the human leukocyte antigen (HLA) system. And this group of genes that regulate the human immune system are also responsible for our unique odor-prints. The variability in the HLA cluster of genes results in one's own personal and particular bouquet.

The MHC complex aids animals in identifying their own offspring and family members. It also helps them assess a potential mate. The vari-ability seen in the MHC complex is not just responsible for that odor-print; it is also linked to a healthy immune system. It's no different in humans: the more variability in your HLA, the better your immune system. Because of this, it was long hypothesized that folks were most attracted to those who had an MHC complex as different from their own as possible—a little sniffable yet unconscious note to lock on to the mate that will help us produce the strongest, healthiest offspring.

Over ten years ago Swiss researchers at the University of Bern undertook what I have now come to think of as the "stinky tee" experiment.[5] The group genotyped forty-nine female students and forty-four male students, looking specifically at their HLA gene sequences. The men were then asked to live "odor-neutral" for a few days, avoiding sexual activity, odor-producing foods, and cigarettes; on two consecutive nights they slept in a T-shirt provided by the researchers. When the T-shirts were returned to the experimenters, the women were asked to sniff six different shirts and rate them for intensity, pleasantness, and sexiness.

The researchers found that HLA mattered: women rated the odor of the men whose HLA systems were more dissimilar from theirs as more pleasant and sexy, compared to the men whose HLA systems were similar to theirs. This trend was reversed if the woman was taking oral contraceptives. The findings led the researchers to conclude that HLA-linked genes do influence female mate choices, supporting the hypothesis that divergent MHC complexes will result in attraction.

A more recent study, led by McClintock, challenged the notion that the more different, the more attractive. The group did their own "smelly tee" test with forty-nine women. In this case the researchers selected a diverse group of male sweaty T-shirt donors, but made sure there was some overlap with common HLA alleles found in the female smellers' families. McClintock and her colleagues discovered that women were able to discriminate the differences in HLA genotypes. However, they did so based on the HLA alleles they inherited from their father. What does this mean, exactly? These women did not prefer odors that were completely different from their own HLA system–influenced odor-print. Rather, they preferred odors that had a couple of alleles in common with dear old Dad. It would seem that women are programmed to prefer a man who retains some genetic similarity to their own paternal ancestry.

"Women need to optimize a choice of mate in hopes of providing the best possible immune system for offspring," said Charles Wysocki, a researcher at the Monell Chemical Senses Center in Philadelphia. "She can do this by choosing a male whose immune system genes are different from hers. But not maximally different, optimally different."

So if at last year's family reunion you found your fourth cousin (twice removed) on your dad's side more attractive than you thought you should, you are not quite as weird as you thought. Your olfactory system was just on the lookout for optimal genetic diversity.[6]

Several online dating companies now offer HLA matching services as part of their programs. For a cool grand, you can have potential matches genotyped to see how closely related your HLA systems are. While doing so probably won't hurt your dating scorecard, there's no guarantee that similar HLA genes make for a happy relationship. And the HLA complex is not the only chemical being released by the body. Wysocki argues that there may be more olfactory cues that can help women suss out optimal partners. They simply haven't been characterized yet.

Sexy as a Boar's Saliva

Though scientists have not officially designated any one chemical as a "pheromone" in humans, they have identified a few in other mammals. The most well-known of these is a compound called androstenone. When female pigs in heat get a whiff of this stuff in a boar's saliva, they immediately assume lordosis, or the arched back ready-to-get-busy position. You can even buy it in a convenient spray can, a product called Boarmate, for all your swine husbandry needs. It is used in pig farms across the world to assist in artificial insemination.

As it so happens, androstenone and a related chemical called androstadienone also reside in human male axillary compounds—that's fancy talk for pit sweat—as well as in their saliva and urine. This inclusion has led some to suggest that these two compounds also work as pheromones in humans. Others, however, contend its girl-getting properties are unique to pigs.

This debate hasn't stopped several companies, most Internet-based, from marketing and selling androstenone as a "pheromone" fragrance. Men can buy their own androstenone, which, according to the sellers, will assist in picking up the ladies. But Wysocki cautioned that we shouldn't believe all the hype. "Pheromones aren't what you read about on some of those Internet sites," he told me. "Human evolution has stepped away from the reflexive response to pheromones we see in moths, rats, or mice. We have a lot of cognitive input in how we respond to situations—there's a lot more going on than pure reflex." He paused for a moment. "We know there is some unconscious processing of human body odor. And there is some evidence to suggest body odor can help us identify individuals we know or perhaps attract us to others. But there is simply no good, reliable experimental evidence to support the claim that some pheromone spray you buy on the Internet is going to help make you more attractive to others."

More than a few who have balked at the price tag for an androstenone cologne made for humans have instead opted to try Boarmate as a personal fragrance. Some have even written about their experiences on the Internet. These tales all seemed to be very positive. Yet I found myself wondering, if these folks were so busy getting busy thanks to this pig pheromone, how they have the time to write so eloquently (and frequently) about their many sexual exploits.

One of the most interesting things about androstenone is that not every human can smell it—and that those who can think it smells either quite pleasant, like vanilla, or absolutely disgusting. You would think that if it were a human pheromone guaranteed to attract the ladies, it would smell yummy to every woman. I was certainly curious about what it might smell like to me. I decided to find out. Though Wysocki told me I could pick up a can of Boarmate at any agricultural supply shop, it was harder to track down than I anticipated. Eventually, after a lot of searching, I found an online store in the United Kingdom willing to ship me a can.

When the package arrived in the mail, I quickly pulled it open to find a small yellow aerosol can emblazoned with a cartoonish red pig. Hardly sexy. Of course it's meant for use with pigs, not guys wanting to pick up hot chicks at the local watering hole. Curled up on my bed, my female cat, Boo Boo, at my feet, I read the application instructions: "Spray BOARMATE™ at the gilt or sow's snout for two seconds from a distance of approximately 60 centimeters, then apply pressure to the back of the sow." If, after you press down on the pig's back, she assumes the lordotic position, she's ready to go. It's time to inseminate. Simple enough.

I suppose I could have just sprayed Boarmate in the air and taken a whiff to satisfy my curiosity. But Boo Boo, blissfully snoozing at the foot of the bed, gave me another idea. Yes, it was a mean idea. A despicable idea, really. But I could hardly push on my own back after application to see if I would go into lordosis. Humans don't do lordosis. But cats do. So I sprayed near poor Boo Boo's nose for two seconds and pushed lightly on her back.

Being a kind and placid animal, the best of pets, she did not rear up and scratch out my eyes for experimenting on her in such a manner. Frankly, I wouldn't have blamed her if she had. She simply sniffed at the spray with wide-open eyes, shook her head with menace, and ran out of the room as fast as her little kitty legs could carry her. But not before I could give a quick push down on her back. It probably doesn't come as a surprise that there wasn't any lordosis. She fought my push every step of the way, mainly because she was trying so desperately to make a break for it. Of course, it might also have had something to do with the fact that she wasn't in heat; the Boarmate instructions said I should wait until the sow was in estrus before spraying. But as she was a cat, not a sow, I hardly felt the need to follow the instructions to the letter. Neither do most humans who use the stuff themselves: they don't spray an unsuspecting girl at a club with androstenone, an act that might get them slapped or arrested; they spritz themselves. And I doubt they bother to wait for estrus either. Still, as I said, it was a mean idea. My quick spray did not result in an animal ready for some hot reproductive action. Instead it just pissed off a normally loving feline who refused to come near me for the rest of the day.

The experiment was a success in one respect, however: it allowed me to smell androstenone for myself. I fall into the category of humans who can smell it. Inhaling deeply, I did not think it smelled like vanilla in the slightest, nor did it smell like garbage. For me, Boarmate was reminiscent of the precursor of deodorant failure. You know how sometimes you get a whiff of what might be your own body stink, so you try to inconspicuously sniff under your underarm to check if you need to wash up? Upon careful reflection, I concluded androstenone smells most like that sweaty antecedent—the almost pit stink—to me. If I were to smell it on a guy who was trying to chat me up, I'm fairly certain I'd try to keep some distance, or maybe walk to the ladies' room to make sure the smell wasn't coming from me.

Why might there be such differences in how people perceive androstenone? It all comes down to a type of olfactory receptor. "Androstenone has been known to have a very different olfactory percept for different people," said Hiroaki Matsunami, a molecular geneticist at Duke University. "We were interested in finding out the genetic basis of that."

Matsunami and his colleagues collected blood from nearly four hundred individuals for genotyping. The group was particularly interested in olfactory receptor genes—that is, receptors in the epithelium of the nose that perceive smells. The study participants were then asked to rate the intensity and valence of more than sixty different smells, including androstenone and its close cousin, androstadienone. The group discovered that one particular receptor gene, called OR7D4, was linked to whether individuals could smell these pheromones. And a particular variation, or polymorphism, of the gene in some individuals was correlated to whether they found the smell pleasant or grody.[7]

So there appears to be a human olfactory receptor that is specific to this chemical, a known pig pheromone. When I asked Matsunami if this work provided evidence for or against human pheromone signaling, he offered caution. "The idea is still very controversial, and there is evidence on both sides," he said. "If my contribution about the variation in these receptors leads to understanding of how these chemicals can have a pheromonal effect in humans, that would be nice to see. But I haven't drawn any conclusions yet."

Matsunami was quick to point out that his study involved one receptor out of hundreds. Not to mention that the chemicals androstenone and androstadienone are only two molecules out of potentially millions. The task ahead—to try to understand how the brain makes sense of all these different chemicals and what effects they may have on behavior—is a daunting one at best.

Pheromones and the Brain

Despite the controversy over whether pheromones can really affect humans, a group of researchers at Sweden's Karolinska Institute, including Ivanka Savic, have used positron emission tomography (PET), a neuroimaging technique, to see how pheromones like androstadienone (AND) and an estrogen-like steroid (EST), a compound found in female urine and more common odors like lavender and cedar, affect cerebral blood flow.[8]

Previous studies have demonstrated that exposure to AND and EST can alter mood, heart and respiratory rate, and skin conductance in humans. They've been shown to do so in a gender-specific manner too. AND does this in females, while EST does so in males. Savic and her colleagues wondered if that effect extended to brain activation. To that end, the researchers scanned the brains of twelve healthy men and twelve healthy women as they smelled AND, EST, and odorless air. The group found that AND and EST activated the anterior hypothalamus in the same gender-specific manner seen in other studies. That is, AND activated the sex and reproduction region of the brain in the women, while the EST lit it up in the men. This kind of sexually dimorphic effect is not seen in common odors. Thus, Savic argues, this is evidence that humans are influenced by pheromone-like signals. "Some argue we cannot perceive pheromones because we do not have a [vomeronasal organ]. But it seems that we can read signals from pheromone-like compounds through the olfactory system," Savic said. "It takes another path, a fairly direct path, from the olfactory mucosa up to the brain."

In a more recent study Savic and her colleagues tested anosmic men, or those who could not smell due to nasal polyps, and normal controls, using a paradigm similar to her first pheromone study. As expected, the men who were unable to smell did not show the EST activation. Savic believes this proves that the olfactory system is necessary and sufficient to process pheromone-like signals in humans.[9]

Savic's work remains controversial. The levels of pheromones that she uses in her studies are orders of magnitude higher than what you'd find in your average human. That high concentration, many argue, nullifies the results. "These are levels one thousand times more concentrated

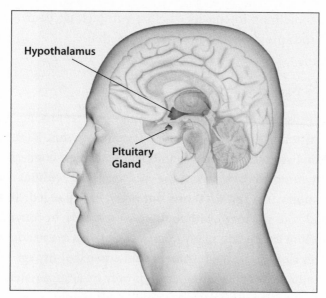

The hypothalamus is activated when humans smell animal pheromones. *Illustration by Dorling Kindersley.*

than what you'd find on the human body," said Wysocki. "If this is supposed to be a human pheromone, you'd expect the same effects at levels found on the human body or even below that level."

Savic acknowledged that the levels used in her experiments may not be the same levels seen in nature, but she argued that her work was still worth further study. Based on her findings, she believes that humans are susceptible to pheromone-like signaling and that future studies will support her hypothesis. But when I asked her about how important they may be to attraction, she paused. "We have not looked at attraction, we have just found these compounds can make physiological changes to the brain," she said. "It is clear that there are ordinary odors we don't consciously perceive targeting specific regions of the brain. That's very compelling and provocative. But there is much more to learn."

"So could attraction be all about pheromones?" I asked.

"No, not at all." She laughed. "Attraction is complex, there are so many factors. Pheromones may be one of those factors. If they are, there are inhibitory connections from other parts of the brain that can augment what those particular odors may do in the brain. But we don't know. It hasn't been properly tested yet."

That's something to consider before you pick up your own can of Boarmate and spray yourself (or your cat) with it.

I Feel the Need for Speed

There is almost nothing more frightening than reentering the dating world as a woman in your midthirties. Trust me on this. Back in the late 1980s, when I started going out with boys, my biggest concern was convincing my mom to drive me to the movie theater without asking too many questions. In a certain sense, not much has changed. Today I still have to deal with my mom, although now I just have to convince her to babysit without asking too many questions. It pains me to admit it, but I am still a bit clueless. Though some would argue that my age makes me a better candidate for the dating pool (if only by appropriately pruning my expectations), I can't help but feel differently.

My friends are pushing me to get back out there, one way or another. One friend, a single dad and busy professional, says speed dating is the way to go. This new paradigm is increasing in popularity, mainly because you don't have to invest all that much time and you are guaranteed to meet more than a dozen "dates" in one evening. Each event involves about twenty women, twenty men, and a good stopwatch. Each singleton spends about four to eight minutes (depending on the event coordinating company), one on one, with each member of the opposite sex. At the end of the evening speed daters have to make a simple decision about each person they met: Would you like to see him or her again? If both the speed dater and his or her persons of interest answer in the affirmative, the event coordinators will send both the corresponding contact information. My friend is a big fan of the setup. He highly recommends it as a fun (but mainly efficient) way of meeting potential dates. Apparently it gets him laid a lot. As it turns out, speed dating is also offering researchers more information about attraction.

In 2004 Eli Finkel, a professor at Northwestern University, took students from a graduate seminar on close relationships, including Paul Eastwick (now a professor at Texas A&M University) on a speed dating field trip. It was a bit of a lark. However, after experiencing it for themselves, Finkel and Eastwick thought it made a good paradigm for study; they were surprised at how much information they took away from only a few minutes of meeting with a potential date. Maybe, they thought, speed dating would provide new insight into what attracts us to other people.

Attraction, like love, has gotten short shrift in the research world, and for the same reasons: it is a complex thing that is hard to define or systematically study. Social psychological research from the 1970s did some looking into the matter. Common wisdom, such as that men appreciate physical attraction more than women and that women are most attracted to a good earner, hail from these studies. Did those results cover the full extent of it? Given the variety of responses I had from just a small group of friends about what they found attractive about their mates, I'd think not.

Finkel and Eastwick wondered the same thing. So the two started holding their own speed dating events, experimental paradigms where members of the opposite sex would meet for four minutes, then judge whether they'd like to meet up with any of their "dates" later. Here, however, each meeting was video-recorded, to be coded for specific behaviors by trained observers later, and all participants received follow-up surveys months after attending.

One of the pair's first findings was a bit of a surprise: that what we think we want in a partner is not usually what we go after. Finkel and Eastwick had a group of speed dating participants fill out a questionnaire about what they were looking for in a partner. In these surveys the old adage that men appreciate physical beauty and women appreciate earning potential held up. However, when those same individuals met dates face-to-face, the ideals were not as important. Rather, both men and women were drawn first to physical attractiveness and then to personality, followed by earning potential. In a real-world situation (or as close as you can get in an experiment), those attraction sex differences we've so long held as universal truths disappeared.[10] Physical attraction trumped all, whether you carried two X chromosomes or an X and a Y.

In an interview with *Newsweek* magazine, Finkel offered this advice to daters based on his study: "Beware the shopping list. When you go into finding a romantic partner, don't have this list of necessary characteristics that you need. Go in with an open mind. Actually meet people face to face. Because you might find yourself surprised by the person you're attracted to."[11]

There's something to be said for that. I've certainly been attracted to a lot of "surprises" over the course of my own dating career. Many I discounted because they did not match my idea of an ideal partner. It makes me wonder if I wasn't a bit too hasty in deciding they weren't for me.

Attraction is not all about good looks; a pleasant conversation is important too. To test the idea, Finkel, Eastwick, and their colleagues looked at language-style matching, or how much individuals matched their conversation to that of their partner orally or in writing, and how it related to attraction. This verbal coordination is something we unconsciously do, at least a little bit, with anyone we speak to, but the researchers wondered if a high level of synchrony might offer clues about what types of people individuals would want to see again.

In an initial study the researchers analyzed forty speed dates for language use. They found that the more similar the two daters' language was, the more likely it was that they would want to meet up again. So far, so good. But might that language-style matching also help predict whether a date or two will progress to a committed relationship? To find out, the researchers analyzed instant messages from committed couples who chatted daily, and compared the level of language-style matching with relationship stability measures gathered using a standardized questionnaire. Three months later the researchers checked back to see if those couples were still together and had them fill out another questionnaire.

The group found that language-style matching was also predictive of relationship stability. People in relationships with high levels of language-style matching were almost twice as likely to still be together when the researchers followed up with them three months later.[12] Apparently conversation, or at least the ability to sync up and get on the same page, mattered.

Granted, these results offer somewhat of a chicken-and-egg problem. Are the couples well suited to each other because they share similar conversational styles? Or do they develop similar conversational styles because they are well suited to each other? It's hard to say. But the researchers believe this is yet another key variable to understanding interpersonal relationships.

Based on the success of Finkel and Eastwick's studies, Jeff Cooper, a postdoctoral fellow formerly of Trinity University in Dublin and now at California Institute of Technology, decided to use speed dating to look at the rewarding aspects of interpersonal attraction. But he tweaked the paradigm a bit so he might be able to add a neuroimaging component. After all, if Helen Fisher's theory is correct and finding a partner is one of the greatest rewards of all, you would expect to see the reward areas of the brain light up when you meet a potential candidate. Certainly Cooper thought it should be so. "We know, behaviorally, there is rein-forcement value in your perception of how others think of you," he told me. "But it's a hard thing to study, especially with neuroimaging. Social rewards activate the brain's reward networks, we know that. But we won-dered if interpersonal attraction, the feeling that someone else likes you, would map in the same way."

Cooper and his colleagues ran six speed dating events and then had participants come to the lab a day or two later to participate in the fMRI portion of the study. There participants were scanned as they viewed a photo of a person they had met followed by that person's yes or no decision about wanting to meet up again. Immediately after that, participants were asked to rate their happiness or unhappiness with the other person's decision about getting together again.

In a preliminary analysis, the group found strong reward system activation—but only when participants received a "yes" response from a person they had also said "yes" to. That correlated with participants' happiness ratings about the other person's decisions. At first glance, these data indicate that a person's liking you is rewarding, but only if you like him or her back.

Cooper and his colleagues wondered if the anticipation of someone else's decision might also affect the brain. So they took a look at the brain activation when the study participants viewed the photo *before* seeing whether or not the person wanted to get together again. Again the pre-liminary analysis showed the reward areas of the brain lighting up like fireworks. Cooper's explanation is that when you like someone enough to give him or her a "yes," you will then anticipate that person's response to you. In the speed dating scenario, it would seem, the reward system activation is active not just for receiving rewards, but also for anticipating them. But getting a "yes" was significantly rewarding only if there was some mutual appreciation going on; otherwise the reward system stayed relatively quiet. "Getting a yes, in other words, varies enormously in reward value depending on how you feel about the person you got it from," said Cooper.[13]

This work is new and ongoing. The researchers plan to do future studies where they scan individuals before a speed dating event. Perhaps there is something in a photo alone that may indicate whether or not you'll end up wanting to go out with another person. Given that physical attractiveness—for both sexes—is a key variable in interpersonal attraction, a picture may be worth more than a thousand words in this scenario. Once the researchers thoroughly analyze all of the data and move on to more complex studies, it's likely they will discover quite a few interesting quirks about what our brains find rewarding when a new person captures our attention. After all, to steal the Facebook line, when it comes to attraction and relationships, it's complicated.

The Take-Home Message

No matter where you look, you can find all manner of advice about how to attract a mate. Magazine covers, dating services, advertisements, fragrance companies—they all want to make us believe there is some key to finding the right person, and they even offer that one special product or service that will help us. The science, however, says it's not quite that simple. There is no attraction smoking gun, whether we are talking about pheromones or a person's earning potential. The only element that repeatedly qualifies as important is physical attraction. That's not such a surprise—but it's a phenomenon that is difficult to study empirically.

"Physical attraction, in both men and women, is a big thing. It's a pretty consistent finding in research studies," said Cooper. "But there is a ton of variation in what people find physically attractive. People just have different tastes in what they like."

Even the ancient Greeks knew beauty was in the eye of the beholder.

Neuroscience hasn't offered us anything more astute than that in the centuries since. People like what they like—cute butts, smoldering eyes, sculpted arms—and what they appreciate in one potential mate may not be so attractive in another. As Cooper said, it's a tough thing to parse out.

I asked Cooper about the challenges of studying attraction from a neuroscientific perspective. "It's hard," he told me. "Social situations are complicated and heterogeneous. There is a lot of variation anytime you make a decision. There's even more when you involve another person in that decision. You see the same thing in speed dating. Even among the people you match with, you aren't going to feel the same about all of them. You may say yes to different people for very different reasons.

"It's a complex situation with a lot of stuff going on," he continued. "Clearly, the things we have to lump together as 'interpersonal attraction' in our work are not always the same things."

"So is it worth studying? Can we learn anything of value about attraction?" I asked.

"Yes, we can. I think it's very interesting to know you are using a lot of the same systems in making a decision about a dating partner as you would in an economic game," he said. "Deciding whether to share five dollars with another person in an economic game overlaps quite a bit at both the computational and neural levels with a person saying, 'Hey, do you want to go out sometime?' That's an interesting thing to learn about humans."

"So attraction is mostly about weighing a bunch of different variables and calculating some output?"

Cooper laughed and tapped his head. "It's all numbers up here at some point."

"Wow, that's so romantic!" I replied sarcastically.

"I think it's also important to know that attraction is highly context-dependent, and it's very fast," he said. "You don't receive any kind of social message in a vacuum. You always receive it against the backdrop of who is giving it to you and what you think or feel about them. And we manage to pull out all the information we need and then make our decisions very, very quickly. Like I said, it's complicated."

I believe him; it is complicated. But even if neuroscience research can't offer me a surefire pheromone or a list of qualities I should seek in a hot date, it has demonstrated that attraction is rewarding, fast, and a little bit different in each and every situation. Knowing, as they say, is half the battle.

"So do you have any advice for the lovelorn? Those who might look to your research and hope for some kind of guidance on how to better attract a mate?"

"Not really," Cooper said. "Except maybe meet more people."

"You mean, get a bigger sample size?"

"Yes, up your N." He laughed. In statistics, N represents sample size. "I don't feel like that's a terribly psychologically inspired answer. But with so much complexity, it certainly can't hurt to get out and meet more people."

Chapter 8

Making Love Last

Monogamy is rare in the animal kingdom; it is estimated that only about 3 percent of mammalian species are monogamous.[1] Humans fall into that small and exclusive category. We have the ability to form strong, lasting emotional bonds with others—bonds that may even last a lifetime.

Helen Fisher believes love is a drive: the drive to find a preferred mate. The human ideal goes further than this utilitarian way of thinking. According to the common marriage vows, we want someone "to have and to hold." We want to meet the person who will inspire us to "forsake all others." We want to promise ourselves to that special someone "until death does us part." You get the idea. Finding love is not enough; we want our love to last. Which, when you consider the matter, is the tricky bit.

Most of us have fallen in love at least a few times in our lives. And not all of us have managed to hang on to that love. Attraction does not guarantee an attachment. Similarly, an attachment does not guarantee a lifelong bond. Several relationship books have come out in the past few years arguing that a better understanding of the body's natural neurochemistry can help solidify a monogamous bond with another person. The premise in these tomes is fairly simple: Understand the brain and you'll understand what it takes to maintain a relationship. They even offer supplements that supposedly provide optimal brain chemistry, whatever that is, and techniques for easy communication to help you in that endeavor. But can neuroscience really present us with any information about what it takes to transform mutual attraction into a loving bond? Can understanding the brain really help us maintain that bond over time?

Monogamy, Vole Style

Remember our friends, the prairie voles? Oh, cute, cuddly *Microtus ochrogaster*. At first glance, you might not think the prairie vole has much to tell us about making love last. This small rodent seems more likely to elicit a squeal from a nine-year-old girl than provide any insight on lasting monogamous relationships. But this species, along with its close cousins, the montane vole (*Microtus montanus*) and meadow vole (*Microtus pennsylvanicus*), have offered scientists a good look into the chemicals in the brain that may underlie an exclusive bond.

Prairie voles are a monogamous species. The risks inherent in prairie living, with limited food and dangerous predators, make it logical to pair up at an early age and stay bonded for life. It is all the more important, out in the wild, to have someone help gather grub and look after the little ones. Prairie voles make pair-bonds immediately after sexual maturity—in fact after a single sexual encounter. It is just your old-fashioned boy-meets-girl, boy-copulates-with-girl-over-a-twenty-four-hour-period, boy-and-girl-live-happily-ever-after type of tale.

In contrast, the montane and meadow voles are the player types. These species of voles, unlike their prairie cousins, live in places with more food and more cover, making it less important to have a second around. Neither of these species forms lasting attachments; rather both males and females have a love-the-one-you're-with mentality. Except they don't really love, more just mate with whoever happens to be nearby and available. Montane and meadow voles are asocial; they do not crave close, personal contact the way prairie voles do. These animals spend only about 5 percent of their time with others. More to the point, they cannot get away quickly enough after mating. Are you familiar with the phrase "coyote ugly," the feeling you get when you wake up after a night of debauchery to find an ugly stranger in bed and are then compelled to gnaw your own arm off to escape undetected? I think there might be some merit in petitioning to change the name to "montane ugly." (Although "meadow ugly" would also apply in this instance, it does not have quite the same ring to it.)

So what might account for these vast differences in mating behaviors in such close species? It all seems to be related to oxytocin and vasopressin. Oxytocin, as discussed in chapter 3, is a hormone produced in the hypothalamus that plays a big part in facilitating labor, birth, and breastfeeding in women. This compound also works as a neurotransmitter, helping to send chemical messages between neurons. It received its nickname, the "cuddle chemical," because of its role in love, pair-bonding, and orgasm. Vasopressin, which is synthesized in the hypothalamus, has multiple responsibilities in the body, including regulating blood pressure and the retention of water. It also works as a neurotransmitter and has been linked to the formation of memories, aggression, and pair-bonding. Here's how oxytocin and vasopressin work their magic on the mating habits of our friends, the prairie voles.

Both male and female prairie voles have a high density of oxytocin receptors, the nucleus accumbens in the brain, making them more social and more likely to prefer a mate than a stranger. Several studies have shown that the more receptors there are in this area, the more friendly an animal will be. Scientists hypothesize that when oxytocin binds with these receptors, it lets loose a flood of dopamine, the "feel good" neurotransmitter that strengthens social relationships. Think of it this way: Sex feels pretty good—very good, when you do it right. When you have sex with your pair-bonded partner, additional dopamine is released through this oxytocin system. And that added dopamine gives the sex that little extra oomph that makes you want to go back to that same partner for more. And more. And probably a little more after that. Some social scientists believe that the reason women are more likely to fall in love with a man they have had sex with—as opposed to one they have just spent time with—is due to this oxytocin-induced dopamine rush.

After prairie voles have mated for twenty-four hours, this extra dopamine causes physical changes to the brain: now sex with anyone else will not be as pleasurable. In fact scientists have observed that the approach of another member of the opposite sex may even cue aggressive behavior. After this first mating happens, prairie voles have no reason to go out for a hamburger when they have steak at home, prepared just the way they like it.[2]

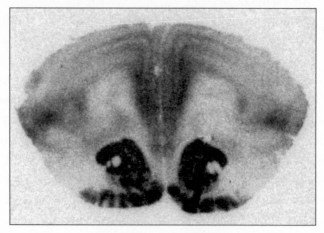

An autoradiogram of the male prairie vole brain. The C-shaped black areas show the density of oxytocin receptors in the nucleus accumbens region. *Photo by Sara M. Freeman, Emory University.*

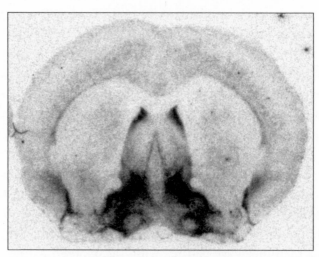

An autoradiogram of the male prairie vole brain. The dark areas highlight the density of vasopressin receptors in the ventral pallidum (the two dark areas toward the bottom). *Photo by Sara M. Freeman, Emory University.*

Prairie voles have a double whammy of chemicals to help them remain monogamous. Along with oxytocin, vasopressin plays a big role in making vole love last, particularly in males. The boys have vasopressin receptors on the ventral pallidum, which also triggers a dopamine storm (and, in turn, strengthens pair-bonds in much the same way that oxytocin does, with the chemical incentive to mate with the familiar partner). Both the ventral pallidum and the nucleus accumbens are part of the reward processing pathway and have been implicated in the human brain when it comes to love and attachment.

It may not sound romantic, but researchers hypothesize that prairie vole monogamy can be chalked up to simple conditioning. Prairie voles mate, the neurochemicals oxytocin and vasopressin are released, they bind to receptors on the reward pathway, and that spectacular dopamine flood takes place. Every time a male prairie vole has sex with his pair-bonded partner, he is getting not one but two extra doses of sugar. With that kind of reward, why would he look elsewhere for love?

But prairie voles' close cousins, the montane and meadow voles, are lacking the density of vasopressin receptors in the ventral pallidum; therefore they do not experience that extra push of dopamine during sex with a particular partner. Without that dopamine, there is no pull for these voles to return to the same female. The only reward is the sex itself—and a montane or meadow vole does not need a specific female for that.

Ultimately the basal ganglia provide a sort of platform for monogamy in prairie voles. With vasopressin receptors present on this brain area, molecular signaling sets these animals up to prefer sex with a pair-bonded partner. When the gene that expresses these receptors is not present, as in the montane and meadow voles, it is more about the booty than any type of bonded relationship. And when it is present, then love, at least as seen in a prairie vole pair-bond, endures—that is, when you use oxytocin, vasopressin, and dopamine-related changes to the brain's reward circuitry as the working definition of *love*.

Dopamine Reception and Maintenance

The neurochemicals oxytocin, vasopressin, and dopamine work together to help voles make a pair-bond. These compounds, mixed together with a little hot sex (as well as a few other chemicals that have yet to be characterized), transform the attraction from a few sniffs of urine into an attachment. What makes that attachment last? What allows it to persevere, even when another sweet-smelling prairie vole saunters into the picture? It appears that it all comes down to the right type of dopamine receptor.

The nucleus accumbens does not only have a high density of oxytocin receptors in monogamous prairie voles, it also houses two types of dopamine receptors: D1 and D2. Previous work with prairie voles suggests that dopamine's pleasurable effects are critical to the formation of a pair-bond. Giving female prairie voles a dopamine agonist, a drug that releases ample dopamine into the brain, allowed them to form a selective pair-bond with a male without the inconvenience of actually having sex. When the researchers blocked the D2 receptor using a drug, it stopped the animals from forming a pair-bond, no matter how many bodily fluids they exchanged.

Given that there are multiple dopamine pathways in the brain, Brandon Aragona, now a neuroscientist at the University of Michigan, and his colleagues at Florida State University wondered if a specific pathway might be responsible for pair-bonding. They soon discovered that the type of receptor in the nucleus accumbens that absorbed all the dopamine released from sex or a drug showed quite different effects depending on the type of vole and his current relationship status.

The group took a pair-bonded male and housed him with a female, either his bonded mate or a stranger. Normally in this scenario, the males will cuddle with their mates and get persnickety with the strangers; the bond means that only an animal's one true love will do, and a different female is viewed as a threat. When Aragona and his colleagues infused the males with a D1 receptor antagonist (a drug that specifically blocked the D1 receptors) directly into the nucleus accumbens, the males were happy to cuddle with whatever female happened to drop in. It was as if their previous pair-bond had never existed. Just a simple infusion and—poof!—"love" disappeared.

Meadow voles naturally have a lot of D1-type receptors residing in their nucleus accumbens. This is true even before they mate. What gives? If D1 is so important to a pair-bond, these guys should be even more faithful than prairie voles. When the researchers blocked the nucleus accumbens D1 receptors in this species of vole, these asocial males all of a sudden became much more affectionate. Although they did not form a pair-bond, the lack of D1 receptors made them much more amenable to company. These differences make it clear that dopamine works in a variety of ways when it comes to vole relationships.

Aragona's work suggests that D1 and D2 receptors have distinct roles when it comes to monogamy. D2-type receptors, in the part of the nucleus accumbens projecting to the ventral pallidum (the brain area implicated in attachment) help animals initially form a pair-bond after that inaugural mating session. The dopamine released during mating is taken up by these receptors and makes physical changes to the brain, resulting in the bond. The male now associates his lady, or her unique odor, as it were, with love.

D1-type receptors appear more important to the maintenance of that pair-bond.[3] After the pair-bond is formed, male prairie voles show an increase of D1 receptors in the nucleus accumbens. Aragona and his colleagues hypothesized that, once voles are bonded, it is the D1 receptors that help ensure a male has eyes only for his one true love. All other females would be ignored or attacked. It is these receptors that are facilitating the double whammy of sugar previously discussed.

It's amazing when you think about it. One chemical can be used in a variety of ways to facilitate these types of social behaviors, and that one chemical makes physical changes to the brain, creating the neural equivalent of a fidelity vow inside the basal ganglia.

Does this differential dopamine receptor action work the same way in other species? Karen Bales, a researcher at the University of California, Davis, and her colleagues looked at D1 receptor binding in titi monkeys (*Callicebus cupreus*), a monogamous primate species. The group used positron emission tomography to measure D1 binding, or how much dopamine was taken up by this specific receptor, in the nucleus accumbens, ventral pallidum, and caudate putamen of adult male titis before and four to eight weeks after forming a pair-bond with a female. The group's preliminary results showed the opposite effect from what was seen in prairie voles: binding potential did not change in the pair-bonded males. In fact it was the unpaired males that showed an increase in D1 receptors in these three brain areas. Why this difference was found requires further study. It may be that titi monkeys have evolved a dopamine system that works a little differently from prairie voles', or there may be a third, unknown variable at work here that has yet to be discovered.[4] Bales plans to follow up on this work with titis, adding in scans months after bonds are formed, to better understand what kinds of changes pair-bonds are making to the male titi brain—and how the dopamine system may mediate them.

Dopamine obviously plays a role in making love last, but what regulates its variety of effects may vary. Aragona suggests that unique distributions of these different dopamine receptor subtypes may underlie species-typical mating strategies.[5] He was talking about different types of voles in this instance. But given Bales's work, it's just as likely that one might observe unique patterns of dopamine receptors across all species, be it prairie voles or titi monkeys or even human beings. Even with similar mating strategies—in this case, a monogamous pair-bond—higher-order mammals may have evolved different dopamine receptor patterns along with their bigger forebrains to facilitate the same sorts of behaviors. It's entirely possible, but more work is needed before we can know for sure.

Oxytocin and Lasting Pair-Bonds

Humans don't want just a long-lasting relationship. They also want a relationship of quality: a loving and beneficial connection. It's assumed that the two go hand in hand, even though we know better. Perhaps understanding the neurochemistry behind a quality pair-bond will offer some answers about how to make love stay.

Cotton-top tamarins (*Saguinus oedipus*), another monogamous breed of monkey, also happen to be cooperative breeders. Moms, dads, and even older siblings all work together to help raise infants. Like humans, they are sexually active not only when they are ovulating; they commonly participate in recreational or "noncontraceptive" sex. Tamarins get down and dirty for the pleasure of the act, not just when the female happens to be fertile. While observing these humanlike tamarin pair-bonds, Charles Snowdon, a zoologist and psychologist at the University of Wisconsin, Madison, noticed that there was quite a bit of variety in relationship quality across different couples. "The animals had very different types of relationships with each other," Snowdon told me. "Some were lovey-dovey with lots of physical contact. These couples are always together, always touching. And they are having lots of sex. Other pairs were almost like random molecules moving around without much regard for each other. They seemed to have very little interest in being together. Honestly, you wouldn't think they were affiliated if you did not know it already."

You see the same thing in human relationships. Some couples, even after years of marriage, remain very physically and emotionally connected. I know one couple, wishing to be referred to herein as Dirk and Lola (probably so they can act it out as some kind of kinky fantasy later), who have been this way since they met in their teen years. More than ten years of marriage later, Dirk and Lola seem to have trouble being in the same room without touching one another. They have sex "as often as they can," which, if they had their druthers, would be at least once a day. They both say they go out of their way to do nice things for each other whenever they can. Although they sometimes fight—they say their arguments can get quite heated—they ultimately remain focused on making each other as happy as possible. Even after years and years together they are still in the honeymoon period of their relationship.

Dirk and Lola can make those around them feel a little jealous over the quality of their connection (not to mention ill—there's only so much happy cuddling folks outside of a pair-bond should have to tolerate). They provide only one side of the relationship quality continuum. On the other side of the spectrum you have couples who act as if they merely endure any time they are required to spend together. You wonder how they've managed to stay in a relationship for so long, given their obviously disconnected lives. These are the people you never see touch—in fact you rarely see them together. They may be technically married, but they don't seem, well, bonded. This is the couple you point at when you say, "I'd rather be alone than be in a relationship like that."

Previous research has shown that oxytocin is released after cuddling and sexual behavior. This led Snowdon and his team to wonder if pairs of tamarins that were more affectionate, engaging in more sex and cuddling with each other, had higher levels of oxytocin than less affectionate couples.

The research group observed the behavior and measured urinary hormone levels in fourteen pairs of tamarins over three weeks. Despite oxytocin's reputation as being a little more important in female pair-bonding, Snowdon did not see an overall difference in oxytocin levels between the sexes. "Within a pair, if one animal had high oxytocin levels, his mate also had high levels," said Snowdon. "If the female had low levels, so did her guy. It was a very close correlation within couples."

There was plenty of variation in oxytocin levels across the couples; it was as varied as the types of affectionate behaviors observed. When the group factored that behavior into the analysis, they found that the contact, grooming, and sexual behaviors explained more than half the variance in oxytocin levels. That is, the more cuddly a couple was, the higher its corresponding oxytocin levels. Conversely, the less affectionate a pair was, the lower its corresponding oxytocin levels.[6]

Snowdon and his team took the analysis a level deeper. Were the same kinds of behaviors driving those oxytocin levels? Or were the males and females getting that boost of the hormone from different kinds of touches? He and his colleagues took a closer look at the data. They found that snuggling and grooming were behind the variance in the females, while for males it came down to how much sex they were having.

In high-oxytocin-level relationships, each member of the pair-bond worked to make sure the other was getting the kind of touching he or she needed. Males initiated the cuddling their mate wanted, and females solicited the sex their mate was after. The findings suggest that there are different behavioral mechanisms behind oxytocin levels and that in the best of relationships, like Dirk and Lola's perhaps, members of pairs make sure to give their partners what they need.

Snowdon's results present a bit of a chicken-and-egg problem. Do pair-bonded couples take care of each other in this way because of their high oxytocin levels? Or does putting their partner first lead to elevated oxytocin? It is difficult to sort out, but when I asked Snowdon what he thinks, he didn't hesitate. "We never got to test this directly," he said. "But I'm inclined to think it's the behavior driving the hormones. Good couples are sensitive to what their partners need. And by giving your partner what they need, you up both of your oxytocin levels."

Snowdon believes this study has direct relevance to humans. "I think we tend to overemphasize the intellectual and emotional sides of relationships and underestimate the physical. Monkeys reconcile with touch and sexual behavior when something perturbs the relationship. It's kissing and making up, really. And there's no reason to think, when human relationships get stressed or there's some kind of separation, that physical contact wouldn't be important to maintaining the bond." He paused for a moment. "It certainly couldn't hurt."

Oxytocin and Humans

Corresponding oxytocin levels have also been found in human couples. Ilanit Gordon, a researcher at Yale University, studies oxytocin levels in relationships. She says the findings are consistent across parents and bonded couples. Apparently oxytocin is important in human relationships too.

"So should I get a man tested for his resting oxytocin level before I agree to date him?" I asked her when we met.

"Studies have shown that levels of oxytocin in romantic partners are correlated. They go up in the first few months of a relationship and then drop to the level we see in parents after that," she said. "Something is happening there. Do you choose someone like you, who has a similar level of oxytocin? Or does coming together sync you up? It could even be both." She paused. "In answer to your question, I don't know that I would test a man before I date him. It's just one marker out of many. For me, I'd check out how sweet someone is before dating him. That would be good enough for me."

"It may not be good enough for some, though. Of course, if I was going to check anyway, I wouldn't even know what to check for. What is a good oxytocin level?"

"Basically people range between fifty and one thousand five hundred picograms per milliliter," she told me.

"That is a huge range," I replied with surprise.

"It is," she agreed with a smile.

I could get myself tested. I could make a potential suitor do the same. But I can't help but feel that there is no point to doing so. Scientists don't know if the oxytocin is driving the behavior or if the behavior is driving the oxytocin. Furthermore it's not just oxytocin at work in the formation or maintenance of a bond. Scientists have identified many other chemicals that also play important roles.

Changes over Time

As discussed in chapter 3, Donatella Marazziti found that romantic love affects one's levels of testosterone, follicle-stimulating hormone (FSH), and cortisol.[7] Being passionately in love also alters nerve growth factor levels in the brain.[8] And all the variations observed during that initial romantic period were no longer present when the couples were retested one to two years later. The chemistry changes as you move from being in the first throes of passionate love to a committed relationship.

Helen Fisher and her colleagues proposed that three distinct yet overlapping systems in the reptilian brain are implicated in love. Sexual attraction has its seat in the hypothalamus. Romantic love resides in the ventral tegmental area and the caudate nucleus, and deep emotional attachment activates the ventral pallidum. All these areas work with dopamine, oxytocin, and vasopressin—they just do it a little differently depending on the type of loving relationship you happen to be in. When Fisher and her colleagues conducted their study, all participants reported feeling passionately in love for one to seventeen months. They were measured only once. One might assume the length of time two people are in a relationship would be one variable that transforms activation from the romantic love system to the one underlying deep attachment.

A group of researchers at the Catholic University of Korea used fMRI to compare cerebral blood flow in five heterosexual couples claiming to be passionately in love, both at the start of their relationship (within one hundred days) and six months later. They used a paradigm similar to Fisher and her colleagues' original romantic love study: study participants were scanned as they passively viewed a photo of their loved one and photos of a close friend of the same sex as their partner, blurred faces, and a gray background without a face. They also filled out a standardized questionnaire to measure just how passionately in love they were.

The researchers found, unsurprisingly, that brain activations did change over time. The study participants were found to be less passionate about their partner after six months. In terms of cerebral blood flow, activation in the caudate nucleus, a reward area that has been implicated in passionate love, was significantly reduced. The authors of the study argued that this reduction in activity demonstrates that romantic love makes dynamic changes as the relationship evolves over time.[9]

But these researchers stopped at six months. What kinds of changes might one find after one year? Ten years? Twenty years? When I asked Fisher if she thinks brain activations will change over time, she answered affirmatively. "Relationships change over time," she said. "I would expect to see that reflected in brain activation too."

Might they change in a way that can be generalized? After all, it may be that not all relationships are doomed to the quiet, warm attachment designated by ventral pallidum activation. Perhaps high-quality relationships, like Dirk and Lola's, are so because the partners somehow manage to stay passionately in love no matter how much time passes.

Recently Fisher and her colleagues looked at the cerebral blood flow of individuals who claimed to still intensely love their partner after decades of togetherness. The group scanned ten women and seven men in the fMRI while they were looking at photos of their long-term significant other, a close long-term friend, a familiar acquaintance, and a not-so-familiar acquaintance. The group found that the brain activation of these long-term lovers closely resembled that of those who were newly in love: there was a lot of overlap in this study and the group's first romantic love study that was published in 2005. Fisher and her colleagues observed that dopamine-rich reward areas like the ventral tegmental area (VTA) and the dorsal striatum were activated in long-term love, as were the globus pallidus, substantia nigra, thalamus, insula, and cingulate cortex. It would appear that it is possible to remain madly in love, even after decades together.

Following the study, the group compared brain activation with scores from a love and relationship questionnaire and found some interesting correlations. The ventral tegmental area and caudate nucleus activation was highly correlated with romantic love scores. Sexual frequency, on the other hand, was linked to hypothalamus and posterior hippocampus activation. The results suggest that love—intense, passionate love—can be sustained over time. It would seem one can, indeed, keep the love alive.[10]

It's a great, inspiring conclusion. But this study, no matter how optimistic it may seem, does not offer us any indication of *why* these couples were able to stay in love after all this time. Though the researchers say there may be specific brain mechanisms to assist in sustaining passionate love, the identities of these mechanisms are yet to be discovered.

Less Vasopressin, More Marital Problems

Recent research done at Sweden's Karolinska Institute suggests that vasopressin has an impact on the quality of relationships. A team of researchers led by Hasse Walum showed that men with a particular variant of the gene thought to be responsible for vasopressin processing are more likely to experience relationship discord when in a monogamous relationship than those with a different genotype.

Walum and his colleagues examined the DNA of hundreds of individuals in long-term, cohabitating relationships (some married, some not, but together for at least five years). In addition the researchers gave the study participants and their significant others questionnaires about the state of their relationship. They then compared those answers to the participants' genetic makeup. The researchers were particularly interested to see if there was any interaction between relationship strength and genes related to vasopressin and its receptors. "We hoped to identify variation in how close these individuals were bonded to their partner," said Walum. "Our main intent was to see if variation in the specific gene that had been shown to be so important for pair-bonding in voles would influence human behavior as well."

Sure enough, that is just what Walum and his colleagues found. Genes are not static; they can evolve and change by interacting with different environmental variables. It is part and parcel of the whole idea of evolution: our genes must change by natural selection, by favoring the mutations that make us more likely to survive in our ever-changing environment. In Walum's study men who had a particular genetic variant, called the 334 variant of the *AVPR1A* gene, were more likely to express dissatisfaction in their relationship. The *AVPR1A* gene is the human equivalent to the so-called monogamy gene in prairie voles; it has been linked to issues with vasopressin receptor expression in the human brain and, intriguingly, autism, a disorder characterized by severe difficulties forming social attachments.[11]

If the individual had a glitch in his DNA that resulted in more than one copy of that varied gene in the chromosome, it actually doubled the chance that he reported a relationship crisis within the past year. A mutation to a single gene—the same gene that seemingly turns monogamy on and off in the prairie and meadow voles—was linked to more problematic monogamous relationships in human males. Interestingly women did not show this same gene-relationship correlation.

As these various lines of neuroscientific research converge, it's clear that oxytocin, vasopressin, and dopamine play important roles both in the formation of a bond and in its maintenance over time. But they aren't the only chemicals involved. As I said in chapter 3, love, in all of its forms, serves up a complex combination of neurochemicals in the brain. Aragona and his lab recently presented new findings demonstrating that a specific type of opioid receptor is also important to maintaining a bond.[12] There are likely more chemicals and receptor subtypes with important roles when it comes to social attachment just waiting to be discovered.

When one considers the unknowns involved in these chemicals and how they influence each other, the problem with so-called love-promoting brain chemistry supplements becomes readily apparent. As these pills modulate oxytocin, testosterone, or vasopressin in the body and the brain (if indeed they even have the power to do so), they may very well be altering other chemical and receptor types that are just as important to social bonds. They may change the body's endogenous production and regulation of these chemicals and receptors over time. Without a better understanding of all the ingredients of this complex neurobiological cocktail, one can never know if those changes are to our benefit or detriment.

Alas, our current understanding of the brain's neurochemistry has not offered all that much to our practical knowledge of how to make your relationship last. There is no one chemical, no magic formula that is responsible for making love stay. Although Snowdon's work with tamarins suggests that a physical connection is important to maintaining a quality bond over time, it's doubtful that this is the only factor involved in human relationships—though, as he said, upping your efforts in that department certainly couldn't hurt.

Chapter 9

———

Mommy (and Daddy) Brain

Have I mentioned how much I love my son? I mean, I really love him—I am head over heels, utterly besotted, crazy in love with that boy. Pick whatever "in love" cliché you like; each and every one fits the way I feel about my kid. If you get me started on all the ways my son is brilliant, charming, and pretty darn perfect, I am likely to continue talking even when your eyes glaze over and you try to force a change of subject. I am that far gone. As I said earlier, my kid may just be "sexy" enough to merit banning—not in the creepy way, but in that irresistible way Nicolas Read quipped about at the first "Is There a Neurobiology of Love?" meeting.[1]

Bruce McEwen, the author of the 1996 Wenner-Gren Symposium meeting report, said he included Read's sexy baby–banning quote to illustrate the power of the mother-child bond and the importance of studying such a phenomenon in a mechanistic way. After I mentioned that the line was one of personal significance to me, McEwen laughed. "That line brings to mind an amazing paper by Craig Ferris," he told me. "He presented pups and cocaine to mother rats and found that the nucleus accumbens was much more activated by the pups than the drug. It's a remarkable finding and justifies the Read quote, in a very practical sense, by showing just how strong the bond between the mother and her pups really is."[2]

And strong it is. I'm not afraid to admit that the way I feel about my son is like nothing else I've ever experienced. It is all-encompassing and unalterable, a true force to be reckoned with. These are feelings that surprise me. Though I had always liked children, I wasn't that girl who wanted to have babies above everything else. I was actually fairly ambivalent about the idea of being a mom. But something changed in me after having the boy. Pretty profoundly too, I might add.

You see these same changes in rats. Virgin rats aren't so keen on the youngsters. In fact they find youngsters pretty aversive and often try to eat or kill them. But something happens after females give birth and then start to nurse offspring: a serious reworking of the reward processing circuitry in the brain that results in moms' finding pups more tantalizing than cocaine. Who knew?

Craig Ferris, a neuroscientist at Northeastern University, noted several studies demonstrating that rat pups are quite the positive reinforcement for moms. Mother rats work hard to be near their pups while they are lactating. They will even, when forced by experimental paradigm, press a bar like crazy to collect pups from a dispenser.

Ferris wondered what was going on in the maternal brain to account for such effects. In an fMRI study he and his colleagues observed that the act of nursing pups lit up the reward processing circuitry in the mothers. Those brain areas, including the olfactory system, nucleus accumbens, ventral tegmental area, cortical amygdala, and hypothalamus, among others, showed a similar pattern of activation in the brains of virgin females after a hit of cocaine. "We knew cocaine lit up this dopamine reward and motivation circuitry in virgin females. And in the moms, pups were lighting up the same circuits," said Ferris. "We wanted to see if there might be a conflict. So we took the moms and gave them cocaine to see what would happen."

Lo and behold, when lactating dams were exposed to cocaine and then scanned in the magnet, there was a *suppression* of brain activity in the reward system. "Not only did the cocaine not activate the motivational system in the moms—it shut it down," said Ferris.

Ferris argues that this period in a dam's life is essential to the survival of the species. A mother rat has to nurse those babies or they aren't going to live. It is a critical behavior—so critical that it has to be more than just pleasurable to the mom. It has to trump any other pleasant distractions that may come her way. After childbirth and lactation the brain is organized in such a way that it will put a negative valence on anything that isn't in the best interest of the pups, even if it is enjoyable, including cocaine. The simple experience of being around the pups is a reinforcer, so to speak, one that helps to change the way the brain's reward circuits operate in these moms.

Those changes result in babies coming first, and everything that might get in the way of caring for those babies second. It reminds me of what my friend, a mom of three, told me when I confessed my marriage was changing for the worse after the birth of my son. She may have had a point when she told me that my intense feelings for my husband might shift to my son. If nothing else, motherhood changed my brain in ways I couldn't have imagined before giving birth. It is hypothesized that the experience of childbirth alters the reward and motivation centers of the brain so mothers not only have an overwhelming love for their children, but also have a strong focus on their children's well-being and proper rearing. What might facilitate these kinds of changes? Wouldn't you know it: animal studies involving both rats, sheep, and prairie voles have demonstrated that our old friends oxytocin and dopamine are involved.

Motherhood Changes Everything

Innately we all, whether we are parents or not, seem to know that there is something unique about the relationship between a mother and her child. Like romantic love, this bond is the stuff of inspirational novels and movies and even gets a fair amount of play in religious lore. Social psychologists have been studying maternal bonds for quite some time, as well as their effect on the physical and mental health of offspring, but neurobiological inquiries into the phenomenon are quite new. The system underlying it, however, is as old as our reptilian brains.

"Motherhood is the biological prototype, the neural and endocrine prototype for all aspects of mammalian sociality," said Sue Carter, one of the pioneers of oxytocin and pair-bond research. "The neurobiological mechanisms involved in the mammalian mother-infant relationship are highly conserved. And I argue that pair-bonds, romantic love, friend-ship, any and all social relationships take root from the circuits and pro-cesses we see in maternal bonds."

Long before researchers like Carter were studying the role of oxytocin in pair-bonds, it was understood that this neuropeptide was involved with motherhood. Physiologically oxytocin has been linked to uterine contractions during birth as well as the milk letdown reflex in lactation, discussed in chapter 3. For a long time scientists believed that was the extent of its role. But soon it became clear that this neuropeptide was also making some kind of cognitive connection between the mind and the body. Consider a nursing mother. She may be resting in the living room, watching a little television or talking with a friend on the telephone, as her infant naps upstairs. If that baby wakes and cries for her, it's not uncommon for milk letdown to occur within a few seconds of Mom hearing the sound. That letdown can occur without any direct touch or interaction with the infant; the baby's cry alone is enough to cause the reaction. This kind of phenomenon made it clear that oxytocin was doing more than just facilitating birth and nursing. It was also playing some kind of psychogenic role, connecting the perception of a cry with an actual bodily response.

In addition, when researchers gave rats, hamsters, and prairie voles infusions of oxytocin, the animals were able to elicit maternal behaviors even if they had not yet given birth. The sum of these results suggests that this neuropeptide has a strong role in maternal behaviors—and, by extension, is related to that special attachment between mother and child.

Receptors and Good Mothering

We already know that lack of oxytocin in rodent models leads to impaired social recognition, problems with spatial memory, and issues with pair-bonds.[3] Unsurprisingly, it works in similar ways with mothers and their children. After all, in order to have a special bond with your offspring, you need to be able to recognize and locate the baby before you even have a chance to care for it. It appears, however, that it is not a particular level of oxytocin in the blood that determines whether rats will have a loving bond with their offspring, but rather the number of oxytocin receptors in different parts of the brain.

As outlined in chapter 4, maternal behavior in rats influences how different genes are expressed later in life, ultimately influencing neural development and the offspring's own behavior down the line. There is a lot of naturally occurring variation in maternal behavior in these species: you'll see high licking and grooming (LG) dams that dote on their babes as well as low-LG dams that fall more into the negligent category. Michael Meaney and Frances Champagne, those epigenetics researchers extraordinaire, wondered what brain changes underlie those differences in maternal behaviors.[4]

Meaney, Champagne, and their colleagues took a close look at the brains of high- and low-LG dams. High-LG animals showed significantly higher oxytocin receptor density all over the brain, especially in a few areas implicated in love, like the medial preoptic area (mPOA), the lateral septum, the central nucleus of the amygdala, the paraventricular nucleus (PVN) of the hypothalamus, and the bed nucleus of the stria terminalis. There was a strong correlation between those receptor levels and the quality of maternal behaviors. To try to better link those oxytocin receptor levels to the behaviors, the group then injected some of the high-LG moms with a compound that blocked the oxytocin receptors. Doing so removed any differences in licking and grooming behaviors between the high- and the low-LG groups. Basically, when oxytocin couldn't be gobbled up by those receptors, every female rat was a low-quality licker.

From the epigenetic standpoint, it would seem that being a high-LG mom helps oxytocin receptor genes create a lot of receptors in certain brain areas. That in turn helps facilitate high-LG behaviors once a mom starts lactating. What about dopamine? Ferris's work demonstrated that lactating lit up dopaminergic reward processing areas in the brain. Mouse studies that knock out dopamine seem to deprive the moms of the ability to properly care for their young. Might dopamine signaling also have something to say about variations in maternal care? In a follow-up study Champagne, Meaney, and their colleagues took a look at the role of dopamine in high- and low-LG behavior.[5]

Using an in vivo technique called voltammetry, a way to measure dopamine activity in live rat mommies, the group found that dopamine neurons fired like crazy in the nucleus accumbens shell, not only as moms licked and groomed their babies, but in anticipation of it. The rate of the activity, in terms of both magnitude and duration, was positively correlated with whether the rat was a high-LG or a low-LG dam. That is, high-LG moms showed significantly higher magnitude and duration of dopamine signaling in the nucleus accumbens shell than their low-LG peers. The researchers argued that this difference might account for the discrepancies observed in the quality of parental care. That higher magnitude and duration of dopamine release in the reward system simply made taking care of pups more pleasurable; hence high-LG behaviors.

In a recent study Meaney's lab connected oxytocin and dopamine in the natural variations observed in maternal care. The ventral tegmental area (VTA) has been implicated in romantic love in a number of studies; it also happens to receive input from the mPOA and PVN, areas that show a higher concentration of oxytocin receptors in high-LG moms. The VTA is connected to the nucleus accumbens, which shows more dopamine activity in the high-LG dams. It's like the brain version of the old song "Dem Bones," but instead of the knee bone connecting to the thighbone, you have the mPOA connected to the VTA, the VTA connected to the nucleus accumbens, and so on. You get the idea. When Meaney and his colleagues directly infused some oxytocin into the VTA, they saw increased dopamine signaling in the nucleus accumbens and higher LG behaviors. When they blocked the oxytocin receptors, they saw lower dopamine activity as well as lower LG behaviors. This study is the first to show that oxytocin directly acts on dopamine release in the brain. So oxytocin release during nursing leads not only to more pleasure but to higher-quality maternal care in high-LG rats.[6]

It's easy to read about these studies and say, "Aha! Good and more loving moms have more oxytocin receptors and more dopamine signaling. We just need to up both of those in all women!" Unfortunately, that line of thinking is an oversimplification.

"I don't really enjoy the good mother–bad mother distinction," Champagne told me. "It doesn't make sense from an evolutionary standpoint. It's not necessarily good or bad to be reared by a low-LG mother. It just prepares you for reproduction and the environment in a different way." As Moshe Szyf, another leading epigeneticist, said, if the genome is the hardware and the epigenome is the software, the mother is the programmer. Her behavior tells the offspring what kind of environment it should expect. Though some of the same systems seem to be at work, both in terms of neural substrate and neurochemicals, licking and grooming during a weeklong nursing period seem hardly akin to what we humans consider maternal love. Might we see the same kind of maternal neurobiology in an animal model that is a little similar to humans?

The Neurobiology of Alloparental Behavior

Prairie voles have a very different parenting model than rats. This species is alloparental; that is, everyone plays a part in raising litters. Seeming to intrinsically know that it takes a village to raise a vole pup, the fathers, brothers, and sisters get in on the act. It's not uncommon for a burrow to house a mother and father as well as several litters of pups of various ages. Young females who have not yet reached sexual maturity may be naturally maternal, grooming babies and retrieving them if they get out of the nest, but this kind of spontaneous parental care occurs in only about half of the girls. It would seem that environment dictates whether these girls will be natural mothers; staying put in the nest after weaning and having Dad around are highly correlated with a prairie vole female's being naturally affectionate.

Oxytocin is also important. A single treatment of oxytocin twenty-four hours after birth can make prairie vole mommies more attentive and loving to their offspring. Given that the density of oxytocin receptors in the nucleus accumbens is profoundly variable across animals, as observed in rats, Larry Young and his colleagues hypothesized that more receptors would be linked to higher alloparental behavior in young females. To test that hypothesis, Young, a neuroscientist at Emory University who studies social bonding, used a viral vector to enhance the production of oxytocin receptors in this brain area. Surprisingly, he found no difference between these animals and controls. It's never quite that simple, is it? Clearly oxytocin is mediating some effects on maternal behavior, but it's unclear how it's doing so in the prairie vole.[7]

Looking to Humans

Rats and prairie voles offer neuroscientists a model, a way to look at how neurochemicals may mediate behavior. But it's not exactly the same behavior we see in humans. People do not have litters of offspring; typically we stick to one child a time. As such, it's not a stretch to imagine that people will naturally have different types of biological parenting strategies, even if we share some neural substrate with rodents. There has been some work done with both sheep and rhesus macaques, animals who show a preferential bond with their young, which demonstrates that oxytocin plays an important role in the recognition of the offspring. Those studies have also found natural variations in maternal care. How oxytocin may be regulated by the environment or other factors, making changes to the brain or influencing other neurochemicals downstream, is still under examination.

Given those limitations, what can we extrapolate from animal models of maternal behavior? It depends on who you ask. When I asked Champagne about how easily the findings translate across species, she was cautiously optimistic. "I think there are certainly clear indicators that, at least at its most basic level, the same processes apply in humans," she said. "Of course, the devil will be in the details when it comes to understanding how it all works."

And there are a lot of details that still need to be discovered. Instead of relying on animal models, many neuroscientists are now trying to study the effects of oxytocin and dopamine directly in human subjects. Ruth Feldman, a psychologist at Bar-Ilan University in Israel, has looked at correlations between oxytocin levels and behavior in women both during pregnancy and after birth. She and her colleagues measured the plasma oxytocin levels of more than sixty pregnant women during their first trimester and third trimester and within one month of giving birth. Then they observed the mothers interacting with their babies, paying careful attention to eye gaze, emotional response, touch, and vocalizations. They also had the mothers fill out an extensive questionnaire about their emotions and behaviors. When Feldman and her colleagues analyzed their data, they found something surprising. A mother's oxytocin level was stable across the pregnancy and after birth. As expected, high oxytocin levels were most predictive of a strong mother-child bond, complete with lots of cooing, attention, and affection, after the child was born, but that level was established long before a mom ever set eyes on her newborn.

Due to the stability of oxytocin levels both during pregnancy and after, a woman's level even early on in her pregnancy can predict her later maternal interactions with her baby. Think about it: this means that even in the first trimester, long before a woman could ever nurse or even gaze at her child, oxytocin is mediating something in the brain that will lead to a certain quality of care. What might that be? No one is quite sure, though it's likely to involve a variety of different processes and neurochemicals, including estrogen, progesterone, and oxytocin.[8]

Neuroimaging and Maternal Love

Several researchers have also undertaken neuroimaging studies of maternal love. Semir Zeki, following on the heels of his romantic love study, compared romantic love to that of a mother for her child. He hypothesized that similar brain circuits would be activated. Once again he relied on visual input. Twenty mothers passively viewed photos of their own infant child (with an age range of nine months to three and a half years), another familiar child of about the same age, the woman's partner, a person she disliked, an unknown child, and an unknown adult. They were instructed to look at the photos and relax while blood flow was measured in the fMRI scanner.[9]

Zeki and his colleagues found that when the women were gazing at photos of their own children there was a significant increase in blood flow in the substantia nigra, the dorsal and ventral striatum, the thalamus, and parts of the prefrontal cortex. Notably, these areas are rich in both oxytocin and vasopressin receptors, as well as dopaminergic neurons—and, as you've noticed, they show some overlap with areas activated in romantic love. The results led Zeki to conclude that love, whether romantic or maternal, is a push-and-pull mechanism. That is, the specific brain activation underlies a softening of our judgment and assessment skills. We view our loved ones, children and romantic partners alike, with rose-colored glasses. When Nicolas Read originally made that crack about "sexy" babies, he was referring to this. Let's face it, one's own baby is pretty damn irresistible. A screaming, pooping, and exhaustion-inducing little machine, perhaps—yet irresistible nonetheless.

I am not exaggerating when I say there is no kid more awesome than mine. My brain tells me it is so. And when my friend Alyson says her kid is the best, she's not wrong either. Even when faced with evidence directly to the contrary, we feel—we *know*—that our kids are fantastic. If we did not have this built-in push-and-pull mechanism, if we did not inherently recognize the "sexiness" of our own offspring, we might choose to not feed them or to leave them on the side of the road when they ask us "Are we there yet?" for the thousandth time on the way to Grandma's house. Couple this neurobiological push on our judgment systems with the pull of dopamine surges in the reward circuitry, and we can create a strong, lasting bond with our children.

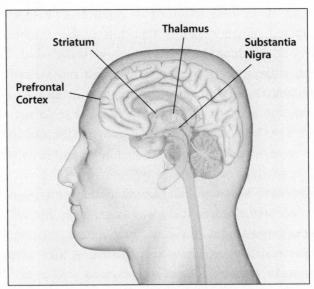

Some of the areas activated in Zeki's maternal love neuroimaging study. *Illustration by Dorling Kindersley.*

Ellen Leibenluft, a scientist at the National Institute of Mental Health, did a similar study with mothers of older children. However, Leibenluft had moms do a simple cognitive task while looking at photos in the fMRI scanner; participants had to indicate whether the photo they were viewing was the same as the one they had viewed earlier. She found a pattern of activation similar to that in the Zeki study, but she also found maternal activation in the anterior paracingulate, posterior cingulate, and superior temporal sulcus. Whether this difference has to do with the children's being older or the fact that the participants were asked to do a cognitive task instead of just looking at photos is unknown. However, it is possible that there are different brain systems involved in dealing with an infant compared to parenting older children.[10]

A Baby's Cry

Photos are all well and good, but babies have other ways of capturing a mother's attention, namely, crying. Might that same sound that could set my breasts a-leaking from across the room also cause a unique pattern of brain activation when heard in the fMRI? James Swain, of Yale University, used fMRI to measure blood flow in nine first-time mothers a few weeks after birth while they listened to a recording of their own baby's cry and another baby's. When moms tuned in to their own babe's cries, Swain and his colleagues found activation in many of the love-related areas, including the basal ganglia, cingulate, amygdala, and insula. When these moms were rescanned three to four months later, the amygdala and insula were no longer activated. Swain hypothesizes that this change in cerebral blood flow may suggest more familiarity. It makes sense; when you are a new parent, a baby's cry can be quite alarming. I spent my first month at home with my son in a bit of a tizzy, looking at every baby-related book and website in hopes of finding some secret instruction manual that would help me decode exactly what he wanted. Over time, as I became more acquainted with his cries, his idiosyncrasies, and his natural schedule, I became less anxious. That, Swain contends, may account for the changes in activation pattern.[11]

Madoka Noriuchi, a researcher at Tokyo Metropolitan University, also looked at maternal neural activation in response to a baby's cry—but her study had a twist. Noriuchi and her colleagues measured cerebral blood flow as moms watched video stimuli, without sound, of an infant crying for his mother as well as a clip in which the infant smiled at her.

These clips, the researchers argued, demonstrated two key infant behaviors that help promote and strengthen the mother-infant bond. The group compared how the mother's brain reacted to her own child versus another as well as between the smiling and crying clips. They found that just watching one's own baby activated the orbitofrontal cortex, the anterior insula, and parts of the putamen. When her own baby was in distress and cried for her, there was additional activation in the caudate nucleus, anterior cingulate, posterior cingulate, thalamus, substantia nigra, and posterior superior temporal sulcus. The response to the crying resulted not only in a different pattern of activation but also in stronger activation. This, the authors argued, suggests the brain is hard at work, taking cues from babies so moms can learn, adapt, and consequently meet the many demands involved in caring for an infant.[12]

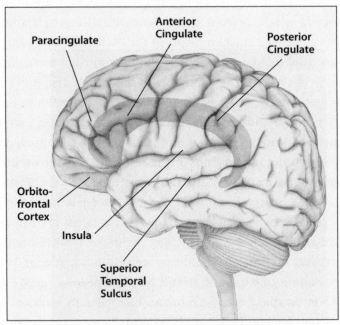

Some of the areas of activation found across different maternal love studies. *Illustration by Dorling Kindersley.*

There has been a significant amount of overlap in the maternal neuroimaging studies, studies looking at romantic love, and studies involving oxytocin and dopamine. With any neuroimaging study in this field, says Stephanie Ortigue, who included these mother love studies in her neuroimaging meta-analysis of love, the question is "So what?" Though we know that many brain areas are activated in a variety of different love-related paradigms, and there has been some replication in findings, it is still hard to pinpoint what exactly these different brain regions are doing and how they are working together to forge the mother-child bond. It is even difficult to know with certainty if maternal attachment is significantly different from any other kind of attachment.

These uncertainties have led some researchers to take a step back. Perhaps before we can make any sense of brain activations involved in maternal love, we need to better understand the kinds of changes motherhood makes to the brain overall. Consider all the bodily alterations that come with pregnancy. The bulging belly and thick, voluminous hair are the obvious ones. Pregnant moms may also scare you with tales of growing a bigger nose and feet. Any mother will tell you it's a whole-body kind of experience. Why wouldn't we see some changes to the brain too?

"The animal literature suggests there are actual structural changes in the brain, especially during the early postpartum period. These are changes where the brain grows bigger," said Pilyoung Kim, a postdoctoral fellow formerly of Yale University. "To be clear, these are local changes—some brain areas experience growth and others remain the same. But the hypothesis is any changes that occur at this time probably play an important role in the development of parenting behavior."

Chief among those behaviors is attachment, or love. But there's more to being a good parent than that. You have to provide, problem-solve, and learn—whether that involves differentiating cries, avoiding dangers, or changing a horrendously poopy diaper without getting any on the only white shirt that still fits you. There is this pervasive notion that new moms are not so smart. The condition is often referred to as "mommy brain," and given the lack of sleep and changing priorities, I can see why the idea is so prevalent. The animal literature, however, does not support the idea that motherhood makes you stupider. Mother rodents navigate mazes faster, capture more prey than virgins do, and do better on a variety of memory and cognitive tasks.

No one, of course, is judging your average rat for accidentally leaving the groceries in the back of the car overnight or losing her keys for the third time this week. So what might truly constitute a "mommy brain"? Kim and her colleagues decided to find out by examining changes in the brains of nineteen moms at a few weeks after giving birth and then three months later. The group found an increase in volume in the prefrontal cortex, the parietal lobes, and many midbrain areas, and when moms had very positive thoughts about their babies, there was an increase in the hypothalamus, amygdala, and substantia nigra. It's a fascinating discovery: positive perception of motherhood mattered in some of those volume changes. That, Kim argues, bolsters the argument that interactions with one's baby, an environmental variable, are just as important as any biological changes that occur alongside pregnancy and childbirth.[13] "Hormonal changes play an important role during the pregnancy, especially toward the end just before and after childbirth," she said. "There's evidence that oxytocin and estrogen levels all go up prior to a child's birth, and that helps the mother's brain change so [she] will perceive the infant as positive and rewarding rather than aversive. But after birth, playing and interacting with the baby also helps with further restructuring."

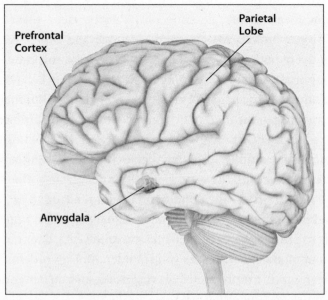

Pilyoung Kim and her colleagues found an increase in volume in the prefrontal cortex, the parietal lobes, and many midbrain areas in mothers a few weeks after giving birth. When moms had positive thoughts about their babies, there was also an increase observed in the hypothalamus, amygdala, and substantia nigra. *Illustration by Dorling Kindersley.*

Kim maintains that both nature and nurture are involved in specific changes to the brain. The brain volume changes suggest an increased capacity for integrating sensory stimuli, improved higher-level cognition, and better ability to enjoy all the dopamine-related rewards of love. But might these changes be necessary only to adapt and care for a newborn infant? When I asked Kim how long these changes last, she paused. "We don't know. It's possible that as long as parenting is a major part of your life, the brain structure remains," she said. "But I can also imagine that once parenting is not a major activity in your life, the brain may go through changes again."

Daddy Brain

So far I have limited the discussion to mothers. It is the primary focus of research in this vein—though, I'll admit, being a mother myself, I have something of a bias. Given that my bodily investment in the birth of my child was a wee bit more time-consuming (not to mention stretch-mark-inducing) than my ex-husband's, I tend to snort when I hear about sympathetic pregnancies and the like. I can't help it. With pregnancy making so many physical changes to my body, it seems perfectly logical that it would alter my brain too. But it is hard to imagine that the brains of expectant fathers would also be affected. Yet new research is discovering that men's brains may also go through some changes once they become fathers (or are made responsible for the care of a child). They may not be the same as those changes that happen in mothers (and I'm tempted to add a snarky "not even close"), but these changes happen nonetheless.

As stated before, prairie voles are alloparental: like moms, dads, sisters, and brothers share in the caregiving. While some virgin females may be a bit *eh* when it comes to pups, sexually immature males get a surge of oxytocin when they are exposed to the little guys. This boost allows them to pair-bond faster when they mate and to become more attentive fathers. If the underlying hypothesis is correct, that oxytocin is mediating some of these changes to the brain, then these lifts likely work some magic on the male brain too.

When Ruth Feldman looked at oxytocin levels across pregnancy and the postpartum period, she did not stop with the moms. She studied the levels of both new moms and new dads soon after birth and then at six months postpartum. And again she had trained observers take a look at parental behavior. The study yielded some interesting findings. First, the oxytocin levels of the fathers looked a heck of a lot like those of their partners during both measurements. Even though birth and lactation helped to push Mom's level up, something was working the same sort of effect in Dad without requiring an epidural. It hardly seems fair.

Second, not only did parents show similar oxytocin levels, but those respective levels were correlated to a gender-specific parenting style. Moms with high oxytocin levels cuddle, coo, and look longingly at their baby. Dads with these levels, however, tend to be more playful and stimulatory, and encourage exploration and interaction with toys. Is this biology or some sort of societal influence? No one can say for sure. It does suggest, however, that oxytocin levels are changing Mom's and Dad's brains in a variety of ways—and in such a manner as to facilitate different kinds of parenting behaviors.[14]

These parenting-style results were further substantiated when Feldman and her colleagues measured oxytocin levels in moms and dads with four- to six-month-old infants after a fifteen-minute play session. The levels were highly correlated across each parent couple. But moms who were über-affectionate with babies showed a significant bump in oxytocin after the play session. Dads, however, showed this increase in the neuropeptide only when they engaged in more physical, stimulatory-type play—you know, the rough-and-tumble type. The group hypothesized that moms and dads have evolved to fulfill different needs in children. And, as was found in epigenetic studies, that early childhood experience has the power to shape neurobiological development when it comes to love and attachment.[15]

It is easy to suggest, especially for us moms, that there is something unique about a maternal bond, that the postpartum brain provides the right substrate for a singular and loving attachment you cannot find anywhere else. That does not, however, diminish the father's role—or that of a grandmother or an adoptive parent or even perhaps a sister or uncle. At the end of the day whoever is offering love and care to a child is likely to be undergoing some changes to the brain that help to cement the bond.

When I spoke with Karen Bales, a neuroscientist at the University of California, Davis, who studied the maternal bond in a variety of species, about whether there was something neurobiologically distinct about a mother's love, she asked me if I knew about the parenting style of titi monkeys (*Callicebus cupreus*), a species of monkey that is monogamous and alloparental. I revealed that I did not. "Essentially," Bales told me, "titi monkeys have a selective attachment to their fathers rather than their mothers. They get a stress response from being separated from their dads; they go to their dads for comfort. I think the idea really is that whomever you grow up having contact comfort with is [the] one you'll have a special attachment to, and that is somehow manifested in the brain. It's the norm in many species to have a special attachment relationship with your mother, but there is no reason to believe that you couldn't have that same kind of relationship with whoever your caregiver happens to be."

I could see that this makes sense. After all, we know that environment is an important variable in developing and maintaining a social bond. When it comes to parenthood, a baby is more than just offspring, it is a learning reinforcement in its own right. One's interactions with an infant play a role in many of the brain changes seen in parenthood. In a pair-bonded couple, brains may change in such a way that moms and dads take on varying parenting strategies. It's just as easy, however, to make the argument that in a situation with two dads (or two moms, or a mom and a grandma, or a dad and an uncle—whatever combination fits a particular family), you'll see a similar division of brain labor, so to speak. Given what we know about the brain's plasticity, there is no reason to think the experience of parenting wouldn't change the brains of adoptive, step-, or foster parents too. The only thing I believe we can say with certainty is that there is still a lot to learn. Honestly, at times the research seems to provide a heck of a lot more questions than answers.

As it stands, the neurobiology of parenting, like that of love and sex, is still in its infancy. There are still many questions to be asked (and answered) about how the brain changes after parenthood and what neurochemical systems may be mediating those changes.

If I'm being honest, no matter what neuroscientists may uncover, they aren't going to change my thinking about my own motherhood experience. No matter how clever or elegant their studies may be, they'll never affect how much I adore my son. As I told you, my feelings are inviolate (and, if Semir Zeki is right, those feelings are supported by a push-and-pull system between my frontal lobes and the brain's reward systems). So whatever the future of this field may bring, I'm going to continue to think my kid is so sexy as to merit banning. I will know in my heart that our bond is like no other. And that's more than enough for me.

———

Chapter 10

Might as Well Face It,
You're Addicted to Love

Turn on the radio at any time of day and chances are you will find a love song playing. It's going to happen whether you are keen on opera, heavy metal, or alt-rock—love is memorialized in song, regardless of genre, more than any other topic. Though any one love song may have its own particular quality, there is a lot of overlap when it comes to the themes explored. One that frequently comes up is the idea of love as an addiction. *Your love is my drug. I can't kick the habit. I don't know why I can't get enough of your love, babe. I don't want no cure. I got to have all your loving. I can't let you go.*[1] I could go on, but you get the idea. There's something about love that tends to make one feel like a crackhead.

Addiction, loosely defined, is the unshakable compulsion to seek out and indulge in a substance like alcohol or drugs, despite the adverse and detrimental consequences that usually go with it. It's a condition that never fully goes away, although it can be treated and may decrease in power over time. Once you give up that substance, you'll suffer physical, perhaps even emotional and mental, withdrawal symptoms. Even if you learn strategies to avoid using, you will still suffer from cravings, and no matter how long you stay clean, you always run a risk of relapse. It's a sticky, messy business.

Though there has been extensive neurobiological study of drug addiction, its etiology, or medical cause, is still not well characterized. What neuroscientists can tell you is that drugs influence the neurotransmitter and receptor systems in the mesocortical limbic system, our reward processing system. Cocaine, for example, blocks the reuptake of dopamine by neurons, giving you that rush of feel-good, albeit manic, energy. The variable (or variables) that causes one person who uses drugs to have just a high old time every now and then while creating a chronic user out of another is still unclear. Somehow the mixture of a genetic predisposition, the social and psychological environment, and repeated tokes of a favored substance leads to the disorder. Makes you wonder—despite all those pop song warblings about love being an addiction, does it actually meet the criteria?

Consider my friend Tasha, who is currently in the throes of love. Tasha's world has started to revolve around her love, eclipsing everything else that was once important to her. Her other personal relationships have suffered since she started dating this guy: she doesn't have much time or patience for anything other than her love. She admits to feeling a physical rush when seeing the object of her love, complete with a racing heart and sweaty palms. When she's away from her love, she misses his presence and his touch. That seems to manifest itself in a physical manner too. She *craves* love. Tasha says the sensation she has when she is with her love is like nothing else on earth. She spends too much money on love, risks her standing at work for love, has argued with family and friends for her love, and has made a few poor life decisions because of love. If she were to be rejected by her love, resulting in his full withdrawal from her life, she would be emotionally despondent, perhaps even physically sick.[2] Tasha's friends have even worried that this single-mindedness may have the power to put her in harm's way. What started as something positive and oh-so-good-feeling has changed in Tasha; this love of hers now often appears as a preoccupation and a collection of somewhat destructive, compulsive behaviors. As you can see, this love has a lot of power over her.

Now go back and replace the word *love* in the previous paragraph with *heroin* or *cocaine*. Love, requited or not, can turn even the most stalwart of us into stereotypical junkies.

Full disclosure: there is no Tasha. I made her up. However, I could put the name of more than a few of my friends in the paragraph above. I would venture to say most of us have felt that overwhelming crazy-in-love feeling for at least one person in our lives, perhaps more than one. We have experienced a love that had the power to make us feel our very best and yet could turn us into the very worst version of ourselves. There's no need to pick on just one of my friends to illustrate the concept.

Sometimes this addiction-like behavior doesn't even involve love per se—just unbelievable, world-rocking sex that ended up holding way too much sway over one's life. In fact I'd wager many of us have also found ourselves inexplicably drawn to a sexual partner whom we did not even like all that much as a person. It happens to the best of us. Even though this situation may include only I-want-to-rip-your-clothes-off lust and bears no resemblance to love, it also has the power to compel you to do things you would not (and often should not) normally do.

Usually these irresistible and overpowering feelings, whether love or lust, mellow over time—hopefully before you do too much damage to your life. The sensation may morph into a strong feeling of attachment or blow up into the unhappiest of endings. We tolerate our friends (and perhaps ourselves) while in the throes of this crazy love because we intrinsically know this too shall pass. At least, we hope so.

But for some, the feeling does not pass. That love itch can never be scratched quite enough. An acquaintance of mine, Kristie, joined a twelve-step recovery program three years ago. Her drug of choice was not heroin or alcohol; she didn't smoke crack or shoot up methamphetamine. No, her compulsive behaviors had everything to do with sex and love. "Before recovery, when I was with a man, it felt like a dream. I was never really present, it was like I was floating above any situation when we were together, feeling physically high," she told me. "When a guy flirts with me, any guy flirts with me, I feel an actual physical high. Like I just got a hit of cocaine. In recovery, we call that 'the intrigue.' And before I got help, I'd feel that hit and do whatever I could, even putting myself in dangerous situations, so I could feel it again."

Currently the *Diagnostic and Statistical Manual of Mental Disorders* (*DSM-IV*) does not include any mention of sexual or love addiction. Kristie diagnosed herself once she recognized the toll her sex addiction was taking on her life. She then sought out a recovery program based on the popular Alcoholics Anonymous model. To date all addictive disorders listed in the *DSM-IV* involve actual substances, not so-called nonsubstances such as gambling, eating, or sex.[3]

"Addiction goes far beyond just drugs," said Wolfram Schultz, a neuroscientist at the University of Cambridge who studies risk and reward processing and how they may be involved in addiction. "In the last twenty years, we've seen other things can become addictive, like sex, food, or even publicity. Some people are addicted to being in the center stage. When you cross that line and *need* that thing you are addicted to, when you get withdrawal symptoms when you can't get it, it's very unhealthy and can be life-endangering."

Individuals like Kristie, who have dealt with a nonsubstance addiction, know it can cause as much physical and emotional damage as one involving Vicodin or vodka. As neurobiological evidence mounts regarding the similarities between all types of addictions, substance or nonsubstance, many are pushing to have the *DSM-IV* category broadened to include them all. One area of focus in better understanding addiction, regardless of its particular basis, is the mesocortical limbic circuitry, including the basal ganglia, and its source of power, the neurotransmitter dopamine.[4]

Developing Addiction

Unexpected rewards lead to the release of dopamine in the basal ganglia, which facilitates learning. The more rewarding a particular stimulus may be, the more likely you will be to see an increase in the particular behavior to get it. We often talk about the basal ganglia as the reward system in the mesocortical limbic circuit. Craig Ferris, the Northeastern University researcher who first demonstrated that rats prefer their own pups to cocaine, told me that it pays to be a little more specific when you talk about this particular circuit. "This pathway is not really a reward system per se," he explained. "It's more a system for motivation, essentially. It's involved with evaluating the risks and the rewards involved with certain behaviors, as well as the predictability of things that may come if you engage in a particular behavior."

Risks *and* rewards: that's important. Just as no good deed goes unpunished, no good reward—be it sex, food, or potential slot machine payout—comes without some risk. One might argue that risk is what gives a particular behavior that little extra oomph, making it exciting enough to want so badly. The mesocortical limbic circuit, involving those dopamine-rich basal ganglia, is not just primed to help you get extra food or laid more often. It is attuned to understanding the risks involved with a particular reward and helping you make the right decisions in order to maximize that reward. When that circuit is out of whack, you may end up with an addiction.[5]

"Addiction is basically a reward process spinning out of control," said Schultz. "Not everyone becomes addicted, that is important. We don't know why some do become addicted and others do not. But in those who do, it's clear there is some flip in the machinery where you need more of a particular reward than you usually get and can't properly assess the risks involved with getting it. That flip is linked to the dopamine system in some way, with the release of too much of it into the brain. Ultimately those high levels of dopamine lead to plasticity in the brain, to changes in the reward processing system that result in no longer being able to make sense of the dopamine signals or how they may relate to the outside world in the proper way."

The long and the short of it is that the reward processing system is compromised by addiction, giving a particular reward a strong value signal, while downplaying the associated risk signal. Over time the compromised reward processing system takes a once positive reward and turns it into a grave negative. Researchers do not know what may cause this wrench in the works (both nature and nurture appear to play a role), but once the switch is thrown, the effects can be devastating.

Love: Addiction's Blueprint?

Nearly ten years ago Thomas Insel, now the director of the National Institute of Mental Health, published a review paper entitled "Is Social Attachment an Addictive Disorder?" Even before love had been correlated with brain areas in the mesocortical limbic pathway in neuroimaging studies, there was quite a bit of conjecture that social attachments might share a common neural substrate with addiction. Studies in animal models showed strong correlations between the neurobiology of mother-infant attachments and pair-bonds with that of drug addiction—and some have suggested that substance abuse could be an attempt to replace the feel-good chemicals usually provided to the brain by bonding with others.[6]

It's a compelling idea. After all, it's hard to make the argument that evolution has naturally selected humans to be susceptible to drug abuse. Instead, perhaps this mesocortical limbic circuit that was developed to promote sex drive and loving attachment was hijacked by drugs, leading to addiction. It's possible. But does that mean love itself is an addiction too?

"Addiction has a negative connotation. We are too quick to say addiction is always a negative thing, and we don't know that for certain," said Schultz. "People who are in love show a number of signs of addiction, including a fascination for a partner, a refocusing of attention, and going to great lengths for a partner. That's clear. And several studies have now shown that love activates the reward system." That would be the same reward processing system implicated in drug addiction.

Schultz maintains that it is easy to acknowledge love as a high-value reward. I think most single folks will tell you that attaining a true and loving life partner is high on their list of personal desires. (I also hypothesize that married individuals may tell you frequent, hot sex is even better.) Neither love nor sex is always rewarding. Both have potential negative—even dangerous—consequences. Some consequences are negative enough that, if we gave them full weight in our decisions, we might never bother to have sex or form attachments.

Consider motherhood. Between morning sickness, stretch marks, and labor pain, popping out a kid is far from sunshine and roses, but it is considered one of the most beautiful experiences a woman can have. Any woman with an episiotomy story (and many of us have really good ones) can tell you that. In response to Pilyoung Kim's fMRI study showing changes to the brain after childbirth, Elizabeth Meyer, a researcher at the University of Richmond, suggested that those brain changes underlie an addiction to motherhood itself. Why is it that women are willing to pop out another kid after the harrowing and painful experience of labor? Perhaps pregnancy and childbirth tweak a mom's mesocortical limbic system—and with it, her assessment of risks and rewards—so she will focus only on the positive aspects of reproducing and consequently continue birthing those babies.

In an interview with *Time* magazine, Meyer, who was pregnant with her second child at the time, was quoted as saying, "Being a mom and also being pregnant right now, it's all very rewarding. If we had to learn from punishing factors, I don't think we'd do it over and over again. The rewards have to outweigh the punishments."[7]

It is likely that our reward system evolved in such a way to compel us to mate and breed in spite of potential negative consequences, to learn to associate our offspring with good experiences instead of the painful (not to mention fattening) ones. It is a great strategy to promote propagation of the species and the healthy raising of young—not so great when that system is commandeered by a regular coke habit.[8]

This evolved system would also have to work its magic on sex and love. It cannot be unique to motherhood; otherwise we would never get to the baby-making portion of the program. We simply cannot afford, from an evolutionary standpoint, to be too put out with a past partner's piss-poor sexual performance or the end of our previous relationship. Over time it's the good stuff that stays with us: the anticipatory heat we may have felt before that sexual disappointment; the good times we shared with our exes; the way we feel when we are physically or emotionally touched by another. If it were the bad moments that remained most memorable, we might not get back up on the horse. Our good memories regarding social attachment must carry more weight than the bad. Our mesocortical limbic system is likely making sure of that. A recent study examining the effects of heartbreak on the brain offers some evidence to support this notion.

Breaking Up Is Hard to Do

Love's more addictive symptoms have a tendency to stick around even if your lover does not. Even after a breakup those in love remain intensely focused (sometimes even more so) on their former partner. The mood swings, obsessive thoughts, personality changes, poor decisions, and lack of self-control also often remain. They may even intensify. The effects of the so-called broken heart are not all that different from the kinds of withdrawal symptoms you may see in a detoxing junkie. Fierce and unrelenting, a lost love can be quite debilitating.

Helen Fisher and her colleagues decided to use fMRI to look at the brain systems activated after rejection in love. The group recruited ten women and five men who had been rejected in love yet could not quite manage to let go of their intended. More to the point, despite all, they wanted the person who rejected them to come back. These folks reported thinking about their sweetheart more than 85 percent of their waking hours. They also admitted to making inappropriate phone calls or visits to their former love, drinking too much, or sobbing for hours on end. As noted in the study's report, participants also showed ambivalence and conflicting emotion when it came to their former lover. They believed their love object was their perfect counterpart, yet also expressed great anger and confusion toward their ex about the way things had ended. Basically they all fit the stereotype of that lovesick fool so commonly found in Dumpsville, U.S.A.

Participants were scanned while they looked at a photo of the lover who rejected them as well as one of a familiar yet emotionally neutral acquaintance. Since they demonstrated such strong emotions toward their former sweethearts, they were asked to do a count-back task, wherein they would mentally count back in increments of seven from a random number in the thousands, in between photo viewings, to clear the emotional decks, as it were. While they looked at photos, they were simply asked to think about events they experienced with the person depicted. These instances tended to be emotionally charged with their rejecter, like a last fight or a poignant romantic weekend together, and fairly boring with their casual acquaintance, like watching television together in a dormitory common room.

Fisher and her colleagues had several specific hypotheses for this study. The two most important were that they'd see some of the same areas light up as observed in the original love study described in chapter 2, particularly the ventral tegmental area, and that they'd see activation in areas involved with drug craving, such as the nucleus accumbens and prefrontal cortex. Sure enough, that's what they found. When they compared the results directly to those from the original love study, they found greater activity in the right nucleus accumbens core, ventral putamen, and ventral pallidum than among those happily in love. This, they argue, is evidence to suggest that the romantic drive is a "goal-oriented motivation state" rather than just an emotion, and a specific form of addiction at that. Just as it brings joy, love can also result in immense sorrow—and even be potentially dangerous.[9]

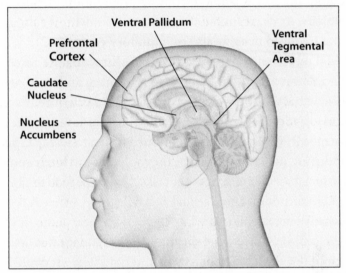

Those rejected in love share patterns of brain activation similar to those who are newly in love as well as to drug addicts. But unlike the other two groups, they also show increased activation in the nucleus accumbens and ventral pallidum. *Illustration by Dorling Kindersley.*

Sexual Addiction

Whether or not the idea of love as an addiction seems plausible to you, most agree that sex definitely has the power to make you its slave. Powerful politicians, famous athletes, and everyday Joes have gambled away integrity, position, and family just because they couldn't resist the siren call of sex. As my friend Kristie can tell you, sexual addiction exacts a hard toll. Work in animal models suggests that certain forms of sexual addiction may be due to problems in the prefrontal cortex.

Lique Coolen, a neuroscientist at the University of Michigan who studies the neurobiology of reward and motivation, was curious about how specific damage to the prefrontal cortex (PFC) may affect sexual behavior, particularly in the areas that link with the mesocortical limbic circuit. She and her colleagues made small lesions in the PFCs of rats, in a specific area that directly connects to the reward circuitry, and then conditioned the rats to associate sex with illness by injecting them with a drug that made them sick after each copulation session. "We've known that the frontal lobe is activated when people are exposed to erotic stimuli for some time," said Coolen. "But despite knowing it was active in sexual activity, its exact function remained unknown. We thought perhaps it was involved in some sort of inhibitory control."

Normal rats would have sex a few times, learn that it caused an icky feeling in their stomach, and stop trying to mate. In fact, Coolen told me, males will actually try to get as far away from females as they can after the association between sex and illness has been made. The animals with damaged PFCs, by contrast, continued trying to mate in spite of the illness-inducing injections. Though they were able to learn and showed normal associative memory in other tasks, this damage to the PFC made them unable to inhibit their sexual behavior even when it resulted in such a negative consequence.[10]

When I asked Coolen what she thought the damage was doing to the reward circuitry, she admitted that she did not know. "It could be some kind of dysfunction in the connections in this circuit or even something that leads to the dysregulation of necessary neurotransmitters. We're not sure," she said. "But it's clear that the prefrontal cortex is important for inhibitory control when it comes to compulsive sexual behavior."

Risky Business

The mesocortical limbic circuit not only processes reward but also has a hand in weighing risk. Most of the studies looking at love or sex as an addiction examine the reward part of this model. But what about the risk? Might there be some change to this system, perhaps to the dopamine signaling in the circuit, that alters one's ability to assess the risks involved with a particular behavior? It would seem so.

Any one neurotransmitter may have an arsenal of potential receptor types. Dopamine is no exception. This chemical's D2 and D1 receptor subtypes have been linked to pair-bond formation and maintenance, respectively, in male prairie voles. However, they have also been linked to addiction. In rat studies stimulating the D2 receptor, the same receptor associated with the formation of a pair-bond, triggers relapse in addicted animals. D1 receptor stimulation, on the other hand, inhibits those cocaine-seeking behaviors. D2 is linked to connecting with another animal or seeking a drug hit, looking for something to satisfy your longing, if you will. D1 is linked to males' avoiding any female but the pair-bonded partner as well as steering clear of cocaine. Here perhaps the animal is satisfied and has no need to look for an outside stimulus, be it love or drug. Although there is still a lot of work to be done in this area, the overlaps are hard to ignore. Taken together these studies support the idea that drugs hijack the brain's natural bonding systems, leading to addiction.[11]

Another type of dopamine receptor, DRD4, is also considered a receptor of interest when it comes to risky business. DRD4 has been linked not only to a variety of sensation-seeking behaviors, but also to addiction. A particular polymorphism of the DRD4 gene called the DRD4 7R+, denoting seven or more repeats of this variant in the genetic code, is a gene of interest in the study of alcoholism, impulsive behavior, ADHD, and anorexia nervosa. It has also been linked to risky sexual behavior.

Of course, there is a lot of variation in sexual behavior. Some people want to get down and dirty at every opportunity; others can take sex or leave it. Some individuals appreciate novelty and a wide range of partners, and others allow themselves sex only within the context of a committed relationship. Given its association with riskier behaviors, Justin Garcia wondered if some of this diversity might be explained by the DRD4 gene. Garcia is an evolutionary biologist at the Laboratory of Evolutionary Anthropology and Health at Binghamton University who studies love and sexual behavior. "This gene is important from an evolutionary standpoint," he said. "It was likely selected for forty or fifty thousand years ago as humans started to push out of Africa and migrate to other parts of the world. It's a gene that is implicated in novelty-seeking, a general sensation-seeking tendency. And it's thought that it gave our ancestors the right kind of motivation to forge that valley and see what's on the other side of the hill."

Might it also be involved with motivations related to forging sexual hills? To find out, Garcia and his colleagues recruited 181 students (118 females and 63 males) to participate in a study. Participants were asked to fill out an extensive, confidential questionnaire about their sexual experience, preferences, and behaviors as well as complete measures that looked at nicotine dependence and impulsive behaviors. The participants also provided a DNA sample by buccal wash; that is, they swished ten milliliters of mouthwash around in their mouth and then returned the used liquid to the researchers for genotyping.[12]

When Garcia and his colleagues compared the genotype of participants, particularly whether they had the 7R+ variety of the *DRD4* gene (24 percent of the overall sample had this particular variation), with their sexual history, they made a few interesting discoveries. First, having 7R+ had no bearing on a person's overall number of sexual partners; however, those with the 7R+ version of the gene did show a higher rate of promiscuity, or number of uncommitted "one-night stand" sexual experiences—nearly twice the rate of those who did not show the 7R+ variation. Though there was not a significantly higher rate of infidelity in the 7R+ group, those who did report being unfaithful had nearly twice the number of sexual partners outside the marriage.

"We've known that dopamine is involved with sexual motivation because of its role in the pleasure and reward circuitry in the brain," said Garcia. "What was interesting to me is that people with this *DRD4* 7R+ type gene did not report having a higher sex drive or higher number of sex partners. It was the context of partners that varied. They went for a more uncommitted, riskier type of sex."

Garcia was careful to point out that this effect is probabilistic rather than deterministic. That is, a higher proportion of those with the 7R+ are motivated to seek risky behaviors when it comes to sex, but having the 7R+ doesn't mean they will.

"Can we really attribute these behavioral differences to just one gene?" I asked him.

"No, there are multiple genes involved with motivating behavior. And there are different environmental factors that will change their expression," said Garcia. "But this is a gene that can explain some of the difference in a behavior that was long thought to be largely culturally defined. That tells us something. That's important."

So Is It an Addiction?

Helen Fisher would tell you that love is no simple emotion, that it is a motivational drive to pursue the greatest of life's rewards: a preferred mate. Because that drive is associated with the dopamine-rich reward circuitry, as well as cortical areas associated with craving, she maintains it also shares qualities with addiction. Several distinct lines of neurobiological research have consistently supported this hypothesis. When I look at Kristie, an admitted sex addict, I see that love and sex can become an addiction in their own right. Drugs may hijack the brain's natural reward and risk processing system, but love and sex can be abused too. What is it about addicts—some kind of genetic predisposition, damage to the reward circuitry or dopamine system—that results in nonsubstance addiction? Neuroscientists don't have a clear answer on that.

It is also important to note that each relationship we have with another person is qualitatively different—often *very* much so. I have had partners I ate, breathed, and slept—I could not get enough of them, despite the fact that we weren't always well-matched. When we parted, I felt as if someone had died. I craved and grieved them. Those relationships felt a lot like what I imagine an addiction would be. By the same token, I have also been involved with men who were pleasant enough companions but did not exactly rock my world. I'm guessing you can say the same. Each relationship is different. These variations raise a good question: Is all love and sex addictive? Or do certain partners, those who bring with them the right cocktail of internal and external chemistry, provide a more addictive state? The latter, intuitively, seems more accurate.

I asked Fisher about this: "I may break up with someone I once loved without a glance backward. Others shatter my heart. Are there any hypotheses about where there is a difference, how different relationships may be affecting the dopamine systems in different ways?"

She was frank in her response. "No one knows. Some people come along at the right time, they fit well into one's personal concept of an ideal partner, and the person falls *madly* in love," she said. "Others aren't quite right in one way or another, and perhaps the dopamine system isn't as activated. But no one knows."

There are many questions left to be answered. Though it would seem that love uses the same neural substrate as substance addictions, it's still unknown what makes us love one person madly and another with calm and ease, what changes are occurring in the brain to cause withdrawal, or even how the dopamine system might be regulated in either case. But for those of us who have experienced a love as irresistible (and destructive) as China White, we know that all those love songs referencing addiction have it right, even if the neuroscience hasn't provided all the details yet.

Chapter 11

Your Cheating Mind

For centuries philosophers, theologians, anthropologists, physicians, and bored spouses all over the world have pondered the nature of monogamy. Is mating with one person over the course of a lifetime *natural*? If it is, then why do so many people go outside their monogamous relationships to, as my friend John so eloquently puts it, "do dirt"?

Statistics about infidelity vary. Do a Google search and you'll find a wide range of numbers. The scientific literature has just as much variability. According to Janis Abrahms Spring and Michael Spring, authors of *After the Affair: Healing the Pain and Rebuilding Trust When a Partner Has Been Unfaithful,* a much-cited source in popular magazine articles, infidelity affects one in every 2.7 couples in the United States: that's 37 percent of couples.[1] Other self-reported polls say 22 percent of men and 14 percent of women are indulging in activities outside their marriage. Many researchers estimate about 25 percent of all married couples or individuals in committed relationships cheat. For the purposes of this chapter, we'll just keep things simple and stick with 22 and 14 percent. Because these surveys are self-reported, many researchers generally assume that the numbers are significantly higher; after all, in many cultures cheating is still very much frowned upon. But if everyone was truly as faithful as he or she claimed, the incidence of sexually transmitted diseases (not to mention divorce) would be much lower.

When I asked Helen Fisher about the prevalence of monogamy (or absence of it, as it were), she told me, "There is not a culture on earth where people don't cheat. I've studied forty-two different cultures across the globe, and you find it in every single one." The question remains: With so many of these same cultures putting a premium on monogamy, why is "dirt" so widespread?

I am sure it is a question that we have all asked ourselves at one time or another, usually while casting sideways glances at the guys. Maybe it has something to do with all those statistics that suggest men cheat more often, but when the topic of infidelity comes up, we seem to automatically assume it is a primarily male pastime. Could it be that monogamy is not a viable option because there is something about the male brain that predisposes men to cheat? It's certainly not a new idea.

An acquaintance of mine—I'll call him Roger—believes that it is his biological imperative to get with as many women as he can. He will tell you straight off the bat that he loves—no, adores—his wife of ten years. In fact he could not live without her. Still he simply cannot resist the siren call of "a little strange" (how he refers to casual sex with strangers) once in a while. To hear Roger talk, usually after a few cocktails, you might think he is bedding a new woman every night. Not so—after all, he has a wife to keep happy. A few times each year Roger uses business trips or boys' nights out to satisfy his desire for that "strange."

Ask a random guy who has cheated on his spouse why he did it and he may tell you that his wife does not want sex as often as he does or that he needs a little more sexual variety in his life. Roger says none of those excuses apply in his own situation. He and his wife have an active sex life. Rather, he believes his genes are to blame for his infidelity; as a male, the thrill of the hunt—the desire to chase and catch a new sexual conquest from time to time—is just a part of his biological makeup that he cannot deny.

Some data from the evolutionary biology field may lend credence to his hypothesis, yet a neuroscientific look into the love lives of cuddly prairie voles—and those exquisite neurochemicals, oxytocin and vasopressin—tells us there's a lot more than just evolution at play when it comes to remaining faithful.

The Evolutionary Perspective

Let's talk numbers. Your garden-variety human male produces about two hundred million sperm per ejaculation after sexual maturity. It can be up to eight hundred million if he hasn't seen any action in a while. There is no "waste not, want not" rule in effect here either. Men can ejaculate as much as they like and their body will just keep producing more sperm.

Women are born with all the eggs they are ever going to have. If you assume a woman starts menstruating around the age of fourteen and will release, on average, one egg each month until she becomes menopausal, you are looking at a rough calculation of twelve eggs per year over thirty-one years of fertility. That amounts to about 372 eggs in a lifetime.

Billions and billions of sperm can be ejaculated in a single month. About four hundred eggs are released over the course of a lifetime. That is a pretty sizable discrepancy.

Since evolutionary biologists argue that deep down we are completely and utterly enslaved by our genes. Human beings are essentially programmed for certain behaviors to help propagate the species. What do our genes want? To be passed down to offspring. Forget life, liberty, and the pursuit of happiness: genes are looking for procreation; they just want to out-reproduce other ancestral lines. Some scientists believe that men, possessors of that inexhaustible sperm arsenal, have been evolutionarily selected over the past 150,000 years to act just like the king of Siam's proverbial honeybee from *The King and I,* fertilizing as many flowers as they can manage to land upon. In order to have the best chance of getting their genes out into the world, it is advantageous for them to spread that seed around and impregnate as many females as possible.

On the other side of the evolutionary gender coin, women are better served by being selective about sexual partners. After all, there is a run on eggs; a girl should not waste one on a bad bet. More important, a woman faces a serious commitment if one of those eggs is fertilized: nine months of pregnancy plus several years of child rearing. It is beneficial for a woman (and her genes) to take her time and find a mate with stellar genes as well as the resources and inclination to help raise a child. "As a basic principle, it's kind of inarguable," said Marlene Zuk, professor of biology at the University of California, Riverside, and the author of *Sexual Selections: What We Can and Can't Learn about Sex from Animals.*[2] "It applies from gophers to dragonflies, and we see, as a general rule, that males are more likely to benefit from having multiple sexual partners while females are not."

I am quite sure Roger would love it if it were this straightforward: Men simply have lots of sperm that need to be spread about the land and are therefore going to—nay, *need* to—cheat. Before any male readers try to use that line of thinking to explain what happened on last month's business trip, Zuk cautions that human behavior is not that simple. Dragonflies have a few neurons to help facilitate behavior. Gophers have a little more going on upstairs, but they are certainly far from being the most complicated mammal in the animal kingdom. Roger may not be in the running for a Nobel Prize any time soon, but he is not a strict slave to his genes either. There's more at work here.

The fairer sex, despite the evolutionary advantage of being selective when it comes to sexual partners, has been known to do no small amount of dirt too. That 14 percent female infidelity statistic? That is nothing to sneeze at. (Remember, that 14 percent consists solely of women *willing* to admit to cheating—the actual number is no doubt higher.) If women should be more selective about their sexual partners due to the scarcity of eggs, they should cheat far less than men. If they already have a good bet at home, cheating seems completely counterproductive from an evolutionary perspective. It would seem that infidelity encompasses more than just evolutionary imperative.

Besides natural selection, what else might factor into the infidelity debate? As it so happens, humans have complex neurobiological structures that underlie behaviors related to romantic love, attachment, and of course sex.

The Cheating Brain

Remember those three distinct brain systems found in neuroimaging studies? Helen Fisher postulates that there are three individual systems for sex, romantic love, and attachment. These systems activate many of the same brain regions, including key areas in the basal ganglia and the frontal lobe. It is that old kaleidoscope again: same parts, different patterns. And that kaleidoscope means it is possible to be both attached to one partner, yet sexually attracted to or even romantically in love with another.

"The way you feel when you are madly in love is different than what you feel after casual sex," Fisher told me. These systems use different neurochemical systems, resulting in different emotional states and behaviors. "Yet there is bound to be some interaction happening between these different brain areas. In a sense, the brain is very well built for both monogamy and cheating."

And how. The frontal cortex likely plays a big role in fidelity. Although all mammals have forebrains, the human frontal lobe is the largest and most complex. Beyond DNA, it is what differentiates us from our primate relatives. The frontal lobe is the seat of what neuroscientists call "executive function"—the place where planning, decision making, metacognition, and other higher cognitive processing and behavior occurs—and it is also implicated in moral judgments and religious belief systems. Considering the fact that it is linked to the basal ganglia circuitry (and often lights up in love-related neuroimaging studies) and contains the most dopamine-sensitive neurons in the entire brain, one would think it also has a say in whether we cheat. It certainly has the right setup to be a candidate for the government of monogamous behaviors, with the frontal lobe processing signals from the romantic love and sex drive systems in the basal ganglia and then acting in an inhibitory fashion when behaviors have the potential to get in the way of long-term attachments.

There is also evidence that damage to this area can change social relationships. In 1848 a railroad worker named Phineas Gage sustained a severe injury to his left frontal lobe. In an explosion gone awry, a metal pole was launched through Gage's eye socket and out through the top of his head. Given the extent of his injury, many were surprised that he lived. More shocking, however, were the changes this injury made to Gage's personality. Before the injury he was considered a jovial and hardworking fellow. His physician, John Martyn Harlow, wrote the following about Gage after the accident:

He is fitful, irreverent, indulging at times in the grossest profan-
ity (which was not previously his custom), manifesting but little
deference for his fellows, impatient of restraint or advice when
it conflicts with his desires, at times pertinaciously obstinate, yet
capricious and vacillating, devising many plans of future opera-
tions, which are no sooner arranged than they are abandoned in
turn for others appearing more feasible.[3]

It was also rumored that this upstanding, moral guy became quite a skirt chaser after his accident. Though due to the state of science at the time—remember, phrenology was the brain science du jour back then—most of the evidence on Gage is anecdotal and somewhat unreliable.

Today clinicians can tell you that damage to the frontal lobe of the brain has been associated with sexual dysfunction, increased sex drive, and so-called sexually deviant behaviors. And recall that Lique Coolen's lab linked damage to the frontal lobe with sexual addiction–type behaviors. Yet what do we know about the role of an *undamaged* frontal lobe in love and sexual behavior?

According to Lucy Brown, Fisher's frequent collaborator at the Albert Einstein College of Medicine, the frontal cortex works with the VTA, ventral pallidum, and nucleus accumbens in the complex dance of love and attachment. It is difficult to tease them apart and definitively say which area is responsible for what. The likely scenario is that all these brain areas work together but perform slightly different functions. "The frontal lobe needs the support of the brain stem. Areas important for decision making are fed by dopamine released by the VTA, and the two areas communicate back and forth," said Brown. "When you are talking about this level of complexity, it's never just one part of the brain."

I can't help but think of Casanova, that poor, sexually unsatisfied rhesus monkey I observed when I visited the Yerkes National Primate Research Center's field station. Even in the nonmonogamous culture of the rhesus monkeys, where he was free to love as he would, he used that forebrain of his to assess a potentially volatile social situation and avoid temptation. He knew what he could lose if he partook of easy sex: his status within the group. Even a monkey understood, despite ample opportunity for sex, that it was in his best interest to steer clear. If a monkey can use that kind of decision making and judgment, it would seem your average human being could too.

This has led scientists to dig a little deeper—to examine the molecular pathways, or the interactions of neurochemicals, enzymes, proteins, and receptors at the level of the neuron, in these brain regions. Is there something about the way that dopamine, vasopressin, and oxytocin work on these parts of our brains that might result in a person's being more or less monogamous? Enlisting the help of our friend the prairie vole, scientists investigate this question.

Your Cheating Voles

As we've already discussed in previous chapters, the basal ganglia provide the platform for monogamy in prairie voles. Vasopressin and oxytocin receptors help these animals learn to prefer sex with a pair-bonded partner rather than a stranger. When the gene that expresses these receptors is not working up to full capacity, as in the montane and meadow voles, it is more about the booty than any type of bonded relationship. If the difference is simply these receptors, finding a way to up them in naturally promiscuous voles should help the animals become more monogamous.

When researchers in Larry Young's laboratory at the Yerkes National Primate Research Center increased the density of vasopressin receptors into the ventral pallidum of male meadow voles, their behavior changed dramatically. Suddenly these once-philandering rodents formed strong partner preferences for a single female. Even without having sex with them, males would bond to one special female—whichever one happened to be nearby. This is a major change for such a lone wolf species.

Similarly, when Young and his colleagues blocked the expression of vasopressin receptors in our faithful prairie vole males, they managed to finally let their inner Don Juans out. No longer able to associate a particular female with that overwhelming rush of dopamine, they became more promiscuous and noncommittal. For male voles, at least, it seems that a single gene, one that determines the density of these oxytocin receptors, may underlie monogamous behaviors.[4] Is it possible that a similar gene governs the same sorts of behaviors in humans?

It's Never Quite That Simple, Is It?

Remember Hasse Walum's *AVPR1A* gene study on relationship satisfaction? When he and his colleagues at the Karolinska Institute examined the DNA of hundreds of individuals who had been in a committed relationship for at least five years, they found that those who had one variant of a vasopressin receptor gene, *AVPR1A,* were more likely to be unhappy in their relationship.[5] Between this study and Young's work on vasopressin, many assumed *AVPR1A* must also have something to say about sexual fidelity. Headlines announcing Walum's results ranged from "Why Men Cheat"[6] to "Infidelity: It's All in the Genes."[7] The assumption was that *AVPR1A* must be responsible if a guy strayed outside a committed relationship. The reporting of results by the media was more than just an oversimplification of Walum's study—it was wrong. Walum did not try to correlate infidelity with *AVPR1A.* He couldn't: the questionnaires did not directly ask the participants if they were unfaithful.

Before you ask the doctor to check your man's blood for the *AVPR1A* variant as some sort of prenuptial test, look more closely. Despite these very interesting results, Walum would be the first to tell you that there is a lot more to a happy marriage than a single gene. He suggests there could be many other reasons these relationships were in trouble. First, there is another effect observed in voles that have had their vasopressin systems altered: aggression and anxiety. Perhaps the relationships surveyed in Walum's study were less happy due to a domestic violence situation or some other kind of mental health issue. Second, some of the participants had children. We newer moms and dads know the extra strains that young children can put on the state of a relationship, ranging from differences in parenting philosophy to the division of responsibilities. It certainly played a role in the demise of my marriage. This variable was not examined by Walum's group. Third, we cannot forget the women in this equation. As they say, it takes two to tango. Though we are more likely to point fingers at the guys when it comes to infidelity, the fact that 14 percent of women self-report having sex outside their marriage can't be ignored. It is entirely possible that some of the relationship angst observed in Walum's study had more to do with female infidelity than an exclusively male DNA variant. The group did look at variations in the *AVPR1A* gene in females but found no significant link back to relationship satisfaction. The effect was found only in males. Without more data, it is difficult to pinpoint an exact cause for the observed interaction.

Dopamine and Fidelity

AVPR1A is not the only gene that has been implicated in fidelity and relationship satisfaction. Recall that Justin Garcia found that the 7R+ variation of the *DRD4* gene, a gene related to risky behavior and addiction, was associated with a higher number of sexual partners outside a committed relationship. And it bears repeating that his study found no significant differences between men and women when it came to this effect.[8]

Again, Garcia emphasizes that a particular genetic variation does not mean you will end up cheating on your partner; individuals who boast the 7R+ variation in the dopamine receptor gene may just be more motivated to seek out riskier sexual behavior, just as they might be more likely to jump out of airplanes, drive fast cars, or eat bizarre ethnic cuisine. "One of the big reasons people give for not cheating is that they don't want to hurt the person they love. That can be enough of a reason for people not to do it," said Garcia. "We're cognitive creatures. We recognize there are consequences to our actions. No matter what our particular genetic makeup may be, we can use our frontal lobes and decide not to cheat."

It is important to note that those evolutionary biologists were right on at least one count: our genes, particularly the ones that influence the vasopressin and dopamine systems, do have something to say about our attachments and perhaps our fidelity too. It is not, however, a deterministic effect. Just because you have a little kink in the genetic code involving one of these neurochemicals, that does not mean you are destined to cheat. It's just not that simple. Furthermore there are a variety of other chemicals and neural pathways, some yet to be discovered, that may also play a role in whether or not we stray.

How Do We Define Monogamy?

To make it more complicated, our paragons of monogamy, the little prairie voles, are hardly as pure as the driven snow. These animals may be socially monogamous, but some of both the males and females are still getting a little on the side. A study by Alexander Ophir, now at Oklahoma State University, found that although prairie voles remained with their pair-bonded partner, not all offspring actually were genetically linked to both parents.[9]

Ophir and his colleagues assessed twenty-six litters of prairie vole cubs in the field to determine paternity. Approximately 80 percent of the litters were sired by the male partner of the mother; the other 20 percent were not genetically related to the bonded male. You guessed it: the prairie vole version of the mailman sneaked in while hubby was away from the nest. And that guy was himself usually pair-bonded to another female. When it comes to vole love, there is a big difference between social and sexual monogamy, not to mention between the lab and an animal's natural setting. It is the same in human beings; there is no biological evidence to suggest that every human, regardless of vasopressin receptor density in the nucleus accumbens or a particular variation in *DRD4,* is naturally monogamous. We may be culturally and socially encouraged to be faithful, but it is unclear how much sway that may have over our biological natures. That fun fact, unfortunately, brings us back to Roger's argument, that cheating may just be an instinctual drive too fascinating to ignore.

Pharmacological manufacturers are putting great stock in vasopressin receptor antagonists; in fact, several are hard at work on a "monogamy" drug based on Young's prairie vole research. Before you ask your doctor for a prescription (or invest in a few thousand shares of Big Pharma stock), remember that genes do not operate in a vacuum. In any monogamous relationship there are all manner of environmental variables related to happiness: how the kids are doing in school, how much money you have and how you spend it, how involved the in-laws are, and how frequently you are having sex. That just about covers what my ex-husband and I would bicker about in an average week. The environment also plays a big part in how our genes, including the *AVPR1A* gene, are expressed.

"If you want to explain all the variation in human pair-bonding, you need to look a lot further than just a gene," said Walum. "I think there is quite a bit of biology involved, but genes can only explain a bit about these behaviors. It is a variety of different factors working together— your genes, your culture, your age, your partner—that creates the true impact. You cannot say one of these things is more important than another."

Monogamous Males Ask for Directions

Ophir, while working with his former advisor, Steve Phelps, also found that the ventral pallidum isn't the only brain area linked to monogamy. The posterior cingulate, a brain region critical to processing spatial information, is also involved. Though prairie voles are monogamous, a small percentage of the species out in the field will never form a pair-bond. They simply wander about, opportunistically hooking up with females when they can. When Ophir and Phelps studied the brains of "residents," monogamous males who do form a bond, and "wanderers," those natural nomads and philanderers, they found no significant varia-tions in the number of vasopressin receptors in the ventral pallidum. They did, however, find significant differences in the number in the pos-terior cingulate, as well as the number of oxytocin receptors in parts of the hippocampus, an area involved with memory.[10] These results led them to conclude that navigating space influences mating tactics—ergo monogamy. In order to successfully breed in the wild, prairie voles need to process not only social information about other animals but also spa-tial information concerning the location of those animals. Having the ability to recognize your mate does not do you much good if you don't also have the ability to find her. It's hard to argue with that logic. Males with lots of vasopressin receptors in the posterior cingulate are more likely to be residents. Low binding on this brain area makes for a wan-derer. Ophir argues that these complementary circuits are critical for an animal to plan its best mating strategy, and they feed into the reward circuitry too. Also, interestingly, it seems the number of receptors in the posterior cingulate is more likely to be passed down to offspring than those in the ventral pallidum, influencing whether future generations will become residents or wanderers themselves.[11]

"So is a wanderer always a wanderer?" I asked Ophir in front of the poster demonstrating these results at a professional conference. "Or will a wanderer one day find the right lovely female and settle down?"

He laughed. "We haven't done that experiment yet. The natural history suggests the voles start off single, wander a bit, find that partner, and settle down. If the partner dies, most will remain single but will still mate with multiple females. Whether there's some kind of reorganization of the brain at each of those steps, I just don't know."

Working Together

It is clear that no one brain area, no one chemical, is more important than the others. And with all this talk about vasopressin and dopamine, you might have forgotten that oxytocin also plays a key role in forming lasting pair-bonds. Shouldn't it have something to say about fidelity as well? Some have hypothesized that though both sexes have both chemicals, oxytocin has more pull on females and vasopressin has a greater influence on males. To get a handle on female fidelity, all one had to do was identify the oxytocin equivalent of the *AVPR1A* gene. To date the studies have not borne out that hypothesis. When Sue Carter compared the effects of oxytocin and vasopressin on partner preference formation and social contact, she found both were necessary in both sexes. In fact, since the two chemicals can bond with the other's receptor—meaning oxytocin can bond with the vasopressin receptor on the neuron and vice versa—it is possible that the two help each other out.[12]

"I thought the data would come out that oxytocin was more relevant in females and vasopressin was more relevant in males. We all did," said Carter. "Instead what we found was that both were important in both sexes. However, males produce more vasopressin in certain important regions of the brain associated with defensive behaviors. It is possible that this higher level of vasopressin helps to explain differences in the way that pair-bonds are expressed in males and females."

The many ways oxytocin and vasopressin may interact in the formation of a bond—and by extension monogamy—is still not known. Evolutionary changes in gene expression can be observed in different physical characteristics in races all over the world. Scientists are only beginning to understand how things we may experience in life, or even while still in the womb, can influence our later monogamous behaviors.

A Question of Epigenetics

Karen Bales, the neurobiologist from the University of California, Davis, studying prairie voles and titi monkeys, examines different developmental effects that may impact how an animal forms social relationships. "The idea is that an animal's early environment, perhaps a stressor or maybe some kind of differential parental care, can possibly have an effect on the oxytocin and vasopressin systems in the brain," she said. "Which, in turn, plays a role in how an animal forms a pair-bond later in life." She hypothesized that exposure to oxytocin or vasopressin at an early age may provide such an epigenetic effect, in which an environmental variable changes the way genes are expressed in the brain. Women are often exposed to pitocin, a synthetic form of oxytocin, to help speed labor and delivery; therefore the discovery of an epigenetic effect taking place in voles would certainly have significant implications for people.

Bales and her colleagues injected a litter of prairie voles with a single dose of oxytocin on the day they were born. Given differences in development, this would equate more or less to the last month of gestation in a human child. The researchers then observed the voles, as well as a control group, while they aged. Once the voles reached sexual maturity, Bales's group noticed an interesting sexually dimorphic effect, which was dependent on the amount of oxytocin to which the pup was exposed.[13] "When males got the oxytocin, that one little injection helped them to pair-bond faster and they showed a higher vasopressin receptor density in the ventral pallidum," she said. But the same dose in females did not change their likelihood of pair-bonding. In fact, with a high enough dose of oxytocin, many of these females seemed to prefer male strangers to the father of their pups.

This implies that infidelity may not be a male biological imperative after all. These epigenetic effects, the right combinations of genes and environmental influences, have the distinct potential to alter the way humans approach monogamy, regardless of gender. "The upshot here is that we're seeing long-term changes in social behavior based on things pups were exposed to," said Bales. "These early experiences are very powerful—just about everything you can do to a baby has the potential to change the brain." So although pharmaceutical companies may be able to create a drug that can increase the number of vasopressin receptors popping up in our brain's reward circuitry, it is very unlikely that this same drug could counteract all the other factors involved in fidelity—even if we were willing to give individuals this drug when they were still babies.

Bales's laboratory is going beyond chemical exposures. These researchers are also examining handling behaviors, living arrangements, and even the effect of different types of parental responsibilities. In a recent study they found that prairie voles that helped raise their siblings had more vasopressin receptors in the amygdala, another part of the brain's reward circuitry and an area of the brain implicated in emotional memory. So even if there is only a single gene at work—which is unlikely—all manner of different environmental variables may have an effect on how that gene is actually expressed in the developing brain.

Knocked Down but Not Out

Though I use my pal Roger to illustrate a stereotypical cheater, he cannot speak for all 22 percent of cheating men (and probably not for the 14 percent of creeping women either). Take a look at the people you know, the books you've read, the movies you've seen; no two cheaters are exactly alike. Love, lust, or simple opportunity may be at work in any given situation. Cheaters may share some qualities, the odd trait or two; they may even share some of the same reasons for straying outside their committed relationship. But their situations are not identical, and neither are their genomes or neurochemical makeup.

While neuroscientists have shown that many factors can influence the way *AVPR1A* may be expressed, what they have not yet looked at is how to factor in individual differences. As I said, no two cheaters are alike, given their unique environments and genes. So no two individuals will express *AVPR1A* in quite the same way either. With so many variables at play, individual differences are crucial to understanding the resulting behavior.

"It's pretty amazing," said Larry Young. "You look at some of these animals and some have high levels of receptors, some have low levels, and then they show different behaviors. It may be a small effect and not a particularly good predictor of behavior. But it's there."

Young and one of his graduate students, Katie Barrett, are studying individual differences in the vasopressin receptor. Using a virus, Barrett is "knocking down" the gene to study its effects on animals. Unlike gene knockout techniques, whereby scientists totally silence a particular gene, knock-down methods allow the gene to be expressed, just not to its full potential. With precision the technique may allow neuroscientists to create groups of animals with different variations of gene expression, and then compare how they act with mates and their offspring.

When I suggested to Young that this made the study of such effects even more complicated, he nodded eagerly. "It muddies the waters, certainly," he said. "But it muddies the waters in a way that is representative of how the waters really are."

Young and Barrett have only just begun this line of research, and Young is careful to point out that even when they finish their work, there will be no hard and fast answer to the question of just who will and won't cheat—there are just too many variables at play. The sum of this work, however, does offer us a lot of interesting information.

"We've shown that a single gene and a variation in that gene can have an impact on something as complex as your relationship with others," he said. "But that impact may be relatively small and not a very powerful predictor of behavior. I wouldn't encourage anyone to go out and buy a test to genotype a prospective partner, because it's going to be wrong most of the time." He paused for a moment and then added, "Though it will be wrong less [often] than chance."

So Back to the Question at Hand

What can our brains tell us about infidelity? Is it neurobiologically hard-wired in men? Maybe in women too? If you were hoping for a simple answer—and perhaps an accompanying genetic test or drug therapy to predict or cure the cheating heart—well, neurobiology cannot offer you anything like that at the moment. It is doubtful that it will ever be able to do such a thing. Converging neuroscientific research does suggest that it is highly improbable that all men are cheaters by nature, but it is possible that *some* men, as well as *some* women, have genetic variants due to early environmental exposures that change their brains in ways to make them more likely to stray.

Historically the finger has been pointed at men when it comes to cheating. And considering that old evolutionary imperative argument, it looks as though we still think in those terms. You know the old adage "Men stray, women stay." But that is just faulty thinking. So if not gender or evolutionary imperative, what does drive a person to cheat on a committed partner? It turns out there is still a lot left to learn, and Zuk says that what science tells us has to be applied cautiously. "It's easy to look at an individual human and say, 'Aha! His genes made him do it,'" she told me. "When you talk about powerful men who cheat on their wives, it's all too easy to say that, evolutionarily, this is just what powerful men do and have always done. That's just a parody of how evolution really works. No individual is held captive by his genes." But a better understanding of the way our genes can be shaped by our environment may one day offer us much more useful information.

"People think of genes as being too absolute. When one talks about a 'cheating' gene, it's like any other gene," said Kim Wallen. "It may create a bias in a certain direction, but it isn't the end-all be-all. Having a certain variation on a cheating gene doesn't mean you will cheat any more than having a certain variation on a gene that regulates height means you'll be tall." Think about it: even if your family is full of gentle giants, those tall genes will not be fully expressed unless you have the right kind of diet and avoid accidents. Genes are not deterministic.

Still, it is unquestionable that our genes do play a role in the way our brains develop and, in turn, how we behave. It is likely that a better understanding of how vasopressin and oxytocin work their voodoo on our brains will give us more insight into the neurobiological underpinnings of monogamy. We cannot ignore that the way those genes may be expressed has quite a bit to do with our environment, both when we are children and later.

What did Roger have to say about all this? After I told him about some of the work being done in the field on monogamy, he sheepishly admitted that his father had cheated on his mother before his parents divorced, was more the wandering type, and also happened to be the original purveyor of Roger's own personal biological imperative theory. "Do you think the way he behaved could have done something to my brain? To my genes and my, what-do-you-call-it, vasopressin system?" he asked.

"Could be," I replied. "But remember, we're not slaves to our genes. There is nothing determined about the biology." I was tempted to tell him that his behavior could also just be a severe lack of judgment on his part that could be handled with a little self-control, but I refrained. Nevertheless, it is clear we still have much to learn about all the ways our genes and brain may influence our behavior when it comes to monogamy. Our early experiences shape how our genes are expressed—which in turn shapes the development of our brains—and influence our behaviors. Even if we could tease apart all the variables and figure out each unique contribution, it is likely we wouldn't be able to make any generalization beyond what we have now. Every individual is different. We will probably never be able to look at a potential partner and know, definitively, whether he or she might one day stray. We just have to go with our gut and hope for the best.

Chapter 12

My Adventures with the O-Team

Some days I feel like I just cannot get away from the female orgasm. No matter where I look, it seems the word, with a big flashing letter O, is following me around. I see it on the covers of popular women's magazines in the supermarket checkout line, along with tips and tricks to make it incredible. I hear it discussed in detail at my book club meetings instead of the month's actual nonfiction selection. It has become the one (and only) topic of conversation for one good friend who has recently started dating a man twelve years her junior. With the recent announcement that a German pharmaceutical company would shut down development of the much anticipated "female Viagra" after a poor showing in clinical trials, the female orgasm is making appearances in both the headline news and late-night talk show jokes.

Never mind that I'm not having all that many orgasms. I'm stuck in the minutiae of an exhausting divorce—not exactly sexy stuff. Getting laid can wait until the lawyers are done conversing, lest I inadvertently give them something else to fuss about. But the rest of the world continues to bring up orgasms at every opportunity.

Ever since Viagra became available in 1998, the world has been clamoring for its female equivalent. In my humble opinion it is the same men who are popping those little blue pills like candy who are actually doing all that demanding. Regardless, both clinicians and pharmaceutical companies seem to be listening. The latest great wet hope was a drug called flibanserin, which acts on two different types of serotonin receptors. It was originally studied for its antidepressant properties, but when patients reported an increase of sexual events while on the drug, a rarity for those on antidepressant medications, Boehringer Ingelheim, flibanserin's developer, decided it might have a higher purpose as a drug to enhance female sexual desire.

According to prevalence studies, research that examines how common particular conditions are within the general population, nearly 10 percent of middle-aged women meet the diagnostic criteria for hypoactive sexual desire disorder (HSDD), the fancy clinical name for low sexual desire. Among the *DSM-IV* criteria for HSDD is feeling some distress about the fact that you are not interested in getting busy. HSDD is commonly found in women about to make the "great change of life." With the success of Viagra in men, female sexual desire drugs seemed to be a seriously untapped market for pharmaceutical companies. After all, who doesn't want to have more sex? That's the assumption, anyway.

Unfortunately, no one has been able to find a good drug treatment for HSDD—and not for lack of trying either. Trials looking at a variety of hormones, including estradiol and testosterone, as well as a variety of other hormonal and drug compounds linked to increased sexual desire, have not shown the desired effects in clinical trials. Pun intended. Despite a lot of hype surrounding the effects of flibanserin (and a somewhat obscene amount of money poured into its development), Boehringer Ingelheim decided to halt its attempts to register the drug as an approved HSDD treatment after it showed no improvement in female sexual desire during trials.

Given Kim Wallen's expertise in hormones, I asked him what he thought of flibanserin's arrested development. He referenced the Melbourne Women's Midlife Health Project, a large longitudinal study headed up by Lorraine Dennerstein that followed nearly five hundred women as they transitioned to menopause. The project took every kind of measure you can imagine, including hormone levels, menstruation details, and a variety of different sexual function questionnaires. With such an extensive study, there was hope that Dennerstein and her colleagues might be able to find some kind of hormonal smoking gun linked to decreased desire in this particular cohort. But no such luck.

"You know what was the biggest factor that [Dennerstein] found in terms of sexuality in postmenopausal women? It was whether they had a new partner. It had nothing to do with hormones," Wallen told me with a sly smile. Simply stated, human sexual motivation is not just a product of hormones or brain chemicals. There's a lot more to it: your relationship status, your age, your culture, and just who you happen to be schtupping at the time.

Dennerstein's findings are reinforced by the results of the most recent National Survey of Sexual Health and Behavior, a large-scale study of sexual and sexual-health behaviors conducted by Indiana University's Center for Sexual Health Promotion. And not just in women transitioning to a nonreproductive hormonal state. This survey asked nearly six thousand individuals, ranging in age from fourteen to ninety-four, about sexual behaviors and the outcome of their most recent sexual event. The ladies had quite a bit to say about both.

While the headlines about the survey results focused on risky sexual behaviors in the baby boomer set and the fact that more individuals are now reportedly giving anal sex a go, the researchers found something that was very interesting to me. It wasn't that men overestimate whether their female sexual partner had an orgasm during their most recent sexual encounter. Most women admit to faking a time or two, and most men believe no woman has ever pretended with them. You do the math. I was not even surprised that women who reported orgasm during their last sexual event didn't stick to straight-up penile-vaginal intercourse, but enjoyed a more varied repertoire of sexual behaviors during each encounter. Another duh, in my opinion. Variety is the spice of life, right? What struck me was that the researchers found that a higher percentage of women reported having an orgasm during their most recent sexual encounter when it was with someone with whom they were *not* involved. Forget commitment. Forget stability. Forget the strength of a long-term relationship where your partner knows you inside and out. Though it is hard to extrapolate exact meaning from survey data, in this case novelty was a better predictor of female orgasm than what every girl supposedly wants: a lasting, loving relationship.[1]

What does this tell us? Simply stated, there is no one thing, no particular hormone level or current relationship status, that sums up a sex drive or one's level of sexual satisfaction. There are a variety of factors that affect female desire—and even though it might not always seem that way, male desire too. Despite the fact that the androcentric view of orgasm (the idea that only a man's big moment really matters in the grand reproductive scheme of things) no longer prevails, the truth is we just don't know all that much about the nature of "normal" sexual desire. We are not even all that well versed on what happens during an orgasm.

"I have a problem with the term and talking about female sexual dysfunctions," said Beverly Whipple, a researcher at Rutgers University who has been studying female sexuality for more than three decades. "There's still a lot we need to know about normative function before we can truly say what's dysfunctional."

She, Barry Komisaruk, and other colleagues at Rutgers may not yet be able to define normative function, but they can tell us a lot about what is happening in the body during an orgasm. And quite a bit of it is happening in the brain.

It's All in Your Head

It is often said that the brain is the most important sex organ. Though it's a cliché, it's true; anyone can tell you that it is hard to reach orgasm when you get distracted thinking about what happened at work earlier in the day or wondering if your butt looks enormous. The brain matters—quite a lot. For the past few decades researchers have been working diligently to understand the physiology of orgasm. A great deal of that focus has remained below the neck, on what's happening with the penis, clitoris, and vagina. However, in recent years neuroimaging studies have shown—surprise, surprise—that the brain plays a huge role in orgasm. You might well think of your brain as one giant gonad. It kind of is, since it's active even when genital stimulation is absent from the experience.

That's right—orgasm is possible with no downstairs action involved. Everyone knows about nocturnal emissions. Wet dreams are among the first topics discussed in sexual education class at school, if your grade school is progressive enough to offer it. These nighttime emissions would be easy to chalk up to reflex, perhaps, or to some kind of developmental quirk. But orgasms during sleep are well-documented across both genders and at a wide range of ages.

I have some personal experience with this. During my pregnancy I had crazy sex dreams. Vivid, bright, and kind of wacky—one of the handful I remember involved clowns and water guns, and I will leave it at that—these dreams usually brought me to an orgasm intense enough to wake me from my sexcapade slumber. My pregnancy books told me this was perfectly normal. In fact research studies have shown that even women without buns in the oven are able to dream their way into the big O, showing increased heart rate, respiration, and vaginal blood flow during the experience. Though this is not a reflexive response, there was stimulation involved. But that stimulation came only from the brain.

It happens not only in dreams. Some individuals with spinal cord injury can still have orgasms despite the fact they have no feeling below the waist. Some epileptics have orgasms as a by-product of their seizures. There have been several cases of brain-damaged individuals who can have orgasms from a veritable cornucopia of nongenital stimulation (nose vibrations, anyone?). You can stimulate the brain chemically and electrically to get off. There are so many ways to bypass the genitals completely and get yourself to that so-called little death. Sure, the penis and vagina are nice accessories to have around. They're just not necessary.

There are even folks who are able to "think off," to just think of something—sometimes things that aren't even all that sexy—to reach orgasm. It sounds unlikely, almost a joke, perhaps a person doing their best *When Harry Met Sally* delicatessen impression. But it is true. I don't know about you, but if I could manage it, I would be hard-pressed to find good reasons to leave the house.

One friend, who would like me to call her Trixie in these pages, is one so blessed. She says "thinking off" is not that explicit, or even that exciting. She often gets through boring conference calls at work by doing it in her office, and no one is ever the wiser. Though it stretched the limits of our friendship a bit, I asked if she might demonstrate. After a couple of margaritas, she agreed.

I expected something fairly grand: a show reminiscent of Sally Albright in all her faking orgasmic glory, maybe with a few hip gyrations and screams thrown in for good measure. It wasn't anything that dramatic. Trixie simply sat back on the couch, closed her eyes, and grew very quiet. I tried not to watch too intently, but there was obviously very little movement. After a few minutes (and a false start or two, when she opened her eyes to see how closely I was monitoring and started to giggle) the only noticeable changes were heavier breathing and clenched hands. For several minutes that was all she wrote. To tell the truth, I was getting a little bored. I would not even have known ecstasy was upon Trixie except that she groaned once, startled me a bit, and then slowly, purposely expelled her breath and relaxed her hands. Now finished, she opened her eyes and sheepishly asked for another margarita.

"That was it?" I asked incredulously. "That was an orgasm?" I worried that my prying eyes had intimidated her a little bit and taken her off-task, so to speak.

"I told you there wasn't much to it," she said, face flushed and a bit embarrassed. "That's it."

"What were you thinking about?" I asked.

She blushed, deepening the already present flush on her cheeks to a Technicolor rose. "Johnny Depp, actually."

"Just Johnny Depp?" I prodded.

"Well, Johnny Depp doing stuff to me. Do I have to be more explicit than that?"

It was not hard to extrapolate from there. But I had to ask just one more thing. Hoping I wasn't going too far, I pressed on. "How does it compare to an orgasm you have masturbating? Or, you know, with a guy?"

"Well, an orgasm is an orgasm," Trixie replied. "It feels good no matter how you get there. But I'd much rather be with a partner." She shrugged. "I just think of this as a more efficient, less messy method of masturbation."

That makes sense. Most of us appreciate a little self-stimulation now and again, whether clitoral, penile, or mental, even though we would prefer to be having sex with someone else. Imagine if you didn't even have to get your hands dirty. Could Trixie's think-off orgasm be the same as one provided by some good old-fashioned masturbation? Neuroimaging studies suggest so; an orgasm lights up the same brain areas no matter how you happen to get there.

Komisaruk and Whipple have found several key brain areas that are active during an orgasm in women: the hypothalamus, amygdala, hippocampus, anterior cingulate cortex, insular cortex, nucleus accumbens, cerebellum, and sensory cortex. Their work has also shown prefrontal cortex activation, while other research done by a group in the Netherlands suggests that an orgasm may actually turn off this part of the brain involved in decision making and executive function.[2] (This is probably due to methodology differences. While Komisaruk's group uses fMRI to study blood flow changes, the Netherlanders use positron emission tomography. The measurements can be a bit slower in this method. Perhaps more important, in the Dutch studies participants were stimulated by a partner. It may be that the frontal cortex activation has something to do with coordinating movement during self-stimulation. No one can say for sure.)

An area of particular interest to researchers has been a part of the anterior hypothalamus called the paraventricular nucleus that produces and secretes oxytocin into the bloodstream, brain, and spinal cord after orgasm. Those secretions, activating many oxytocin receptors in the reward centers and resulting in a massive release of dopamine, may have something to do with why orgasm feels so damn good. These areas are charged whether you are masturbating, have a partner help you with some manual stimulation, or are "thinking off." As Trixie says, it would seem an orgasm is an orgasm is an orgasm, no matter how you play it.[3]

What about brain activation in men? When I told my friend Sarah that I was doing research into this particular phenomenon, she kidded, "Is there anything to study? One might hypothesize the male brain completely dematerializes during ejaculation." Though I am sure there are many women who would raise a sound "Hear, hear!" in response to that (and perhaps a few men would shrug in reluctant agreement), several studies looking at cerebral blood flow in men during arousal and orgasm have found that there is something going on in the brain.

A group at Stanford University compared brain activation and penile turgidity (that's fancy talk for erection) in healthy heterosexual males while they were viewing videos of sports, erotic material, or relaxing scenes. I can only imagine the posters used to recruit participants, maybe something like "Do you like sports? Do you like porn? Would you like a picture of your brain?" Somehow I don't think they had a hard time finding young men interested in participating in this study. However they got participants into the magnet, the researchers found strong activation, correlated with a good erection, in the claustrum, the left side of the basal ganglia that includes the caudate and putamen, the hypothalamus, areas in the middle occipital and temporal gyri, the cingulate cortex, and sensorimotor regions when the men were viewing the erotic videos. Hypothalamus and basal ganglia activation has been implicated in sexual arousal and erection. That these areas, which are involved in oxytocin and dopamine, saw more blood flow was not much of a surprise. The other activations required a little bit of explanation. The claustrum activation coupled with the occipital and temporal areas suggests that participants were both recognizing (and perhaps facilitating) their erections as well as mentally translating whatever was happening on screen in the erotic video to something that might happen to them personally. That, however, is where the results stop. The group did not continue the measurements to orgasm. Perhaps the researchers couldn't find anyone to clean up the fMRI once the participants were finished. And, sadly, there were no mentions of cerebral blood flow results (or erections) during the sports videos.[4]

Janniko Georgiadis, a leading Dutch researcher interested in orgasm, used positron emission tomography to study brain activation during actual ejaculation. He and his colleagues measured cerebral blood flow while each participant was stimulated to orgasm by a partner. The group found activations deep in the cerebellum, anterior vermis, pons, and ventrolateral thalamus, which are all areas located near the brain stem and implicated in movement. They argued that previous findings of basal ganglia and neocortex activation were signs of arousal as opposed to orgasm itself. Thus the cerebellum and related blood flow can be attributed to the muscle contractions linked to the physical act of ejaculation. Their most important finding, they argued, was the ejaculation-related deactivation across the prefrontal cortex. Perhaps Sarah's idea that the male brain, at least the prefrontal level, dematerializes during orgasm was not so far off the mark.[5]

In a more recent study Georgiadis and his colleagues compared orgasm-related cerebral blood flow in both men and women. The group

concluded that the two sexes differed in activation during genital stimulation, but not in orgasm. Given the differences in basic anatomical
equipment, this finding is not exactly a big surprise. One usually handles
a penis a little differently from a clitoris—one would hope so, at least.
Despite these differences observed in stimulation, there was quite a bit
of overlap in activation between the two groups during the big O. Once
again the group reported a lack of activation in the prefrontal cortex, in
both men and women.[6]

Even with their differences, these studies have demonstrated that
orgasm has a distinct brain signature. Do some of these areas seem
familiar? You've seen the hypothalamus and basal ganglia also appear in
neuroimaging studies of romantic love. As Helen Fisher said, it is like a
kaleidoscope: the pattern can change depending on the circumstances.
Once again the brain proves itself to be, as my professor put it, the ultimate recycler. No redundancies here.

So What?

As Stephanie Ortigue said, knowing what lights up in a brain during arousal and orgasm can tell us only so much. Komisaruk and his
colleagues would like to take things a step further. fMRI has evolved
since their initial work in this area. Like a time-lapse camera, new paradigms can follow the brain activation as the orgasm happens so that
the researchers can map its path in real time. But the brain moves fast,
perhaps even faster than current technology can measure. Nonetheless,
Komisaruk aims to get a better idea of how these brain areas interact
during orgasm by studying the order of their activation. "We want to
see if there's some kind of pattern," he told me. "Orgasm is a perfect case
study for seeing how activity can build up in a crescendo and lead to a
climax, a release in the brain. We can apply a constant, continuous stimulation and hopefully see, minute by minute, as different brain regions
are recruited into activation."

Where does the orgasm begin, continue, and end? Komisaruk
guesses the pattern of activation starts in the sensory cortex, an area
called the paracentral lobule, where the brain registers and processes
the genitals' response to stimulation; moves to the PVN, which releases

all that good oxytocin; and ends at the nucleus accumbens and its subsequent dopamine rush. Whether this is the way it works in all cases requires further confirmation.

Hoping to figure out the exact map of activation, Komisaruk and his colleagues scanned a few folks and compiled their initial findings for a presentation poster at one of the biggest neuroscience conferences in the world, sponsored by the Society for Neuroscience. To prepare for this event they quickly gathered participants to create a movie documenting this real-time activation. I volunteered my brain—as well as my naughty bits—to be a part of it.

Performance Anxiety, fMRI style

If you ever want to make even the most cosmopolitan of your friends speechless, telling them you have volunteered to travel to Newark, New Jersey, so you can masturbate to orgasm in an fMRI is a great way to start. Once they overcome the shock, chances are they will start to ask questions. A *lot* of questions. Most I was able to answer. To start, no, I'm not kidding, I'm really going to do it. Really, it is not a joke, I promise. Yes, I will be in the scanner, the same sort of claustrophobic tube you got your knee scanned in that one time. Yes, I know it is a very tight fit. Loud too. Yes, I'll be self-stimulating. How? Clitorally, to be exact, until I reach orgasm. Will I use a vibrator? No, most vibrators have metal, which is a no-no in the magnet. I'll have to rely on my own hands to get the job done. Yes, technically people will be watching—just the scientists who are running the study, I think. But I will be draped for modesty and the only thing they will really be observing, besides my brain on the computer screen, is my hand to signal when ecstasy is upon me. Both Komisaruk and his colleague, Nan Wise, have explained the whole process in detail to me. No, I am not sure I'll actually be able to do it. But, as instructed, I have been practicing at home and will give it my best shot. It seemed that I was going through the same spiel over and over again. Between Wise's careful instructions and my repeated parroting, I felt I knew the procedure backward and forward. Or so I thought.

It occurred to me only the night before I was due to be scanned that I had forgotten to ask the most important question of all: What do I wear to this session? Neither Emily Post nor my most recent issue of *Cosmo* could tell me the proper dress code for self-stimulating to orgasm in an fMRI scanner. And my previous experience in other neuroimaging studies was no help. I had automatically packed some yoga pants and a tank top, thinking of comfort in a confined space, not ease of access to my nether regions. Panicked, I picked up the phone to call Wise.

"What does one usually wear for this sort of thing?" I asked.

"I always suggest a loose-fitting dress with no panties. That's what I wear." Wise, a former sex therapist turned neuroscience graduate student, always pilots the fMRI studies before other participants are brought in. It was old hat to her. "Something loose and comfortable and easy to get into is best."

The only dress I packed was meant for your more typical seduction situations, not fMRI scanning. It may provide easy access to my downstairs bits, but the big metal zipper up the side means it is inappropriate for the magnet. "I'm sorry," I apologized. "I didn't know. I didn't think to pack anything like that."

"That's fine, that's fine," said Wise. "We have access to some hospital johnnies. You can wear one or two of those to cover up."

The thought of a thin, backless nightgown (or two) initiated a growing feeling of performance anxiety. I couldn't help thinking of the confined space, limited movement, loud clanking noises, and me in a hospital johnny. Though I am not the type of girl who needs to light candles, don lingerie, and crank up the Barry White in order to satisfy myself, I do need a little bit of mood to get things going. I was beginning to worry. Would I be able to find any inspiration to explore Ladytown in the kind of setup I was facing?

The next morning, when I arrived at Rutgers University's Smith Hall, a dark 1970s-style building in the middle of the Newark campus, I was in a bit of a panic. Despite spending an hour or two trying to concoct some kind of sexy fantasy about lab coats and confined spaces the previous night, I was still afraid that when push came to shove, I would not be able to reach orgasm.

I recognized Nan Wise immediately; our phone calls had already made her out to be the quintessential mother type, ever ready to nurture and comfort. She wanted to make sure I had eaten a good breakfast, was already asking for my lunch order, and simultaneously reassured me that the scan would be a piece of cake. We walked upstairs to an older office to prepare for my scanning session later in the day. "O-Team Headquarters" was written in big green letters on a side wall whiteboard. I wondered aloud if the lab had made T-shirts yet. "No, not yet," said Wise. "But that's a great idea!"

Waiting for me there was Komisaruk, compact and dapper. His easygoing manner also had a calming effect. Good thing too—the first order of business was to fit me for a head mask, a sort of modern Count of Monte Cristo–type restraint system made of tight plastic mesh. White and blue, the contraption was part low-budget bondage porn prop and part clinical radiation treatment kit. But it was not meant to be pretty. Rather, it was needed to keep my head as still as possible during the scan. Once we started the scan, it would be screwed directly to the scanner bed, meaning that I would be unable to get into or out of the fMRI tube without assistance. As I lay down on a table for my fitting, I tried to casually ask whether any other participants had difficulties reaching orgasm during their studies. I figured a conversation would help distract me from the heated wet plastic he was about to place over my face.

Being fitted for my "sexy" head mask for the fMRI orgasm study. *Photo by the author.*

"A few. But not too many." He pressed the plastic, warm and malleable, across my face and ears, making sure it would harden in an exact contour of my head. "No, we really don't have too much of that. Maybe two or three tops in all the people we've scanned."

Ah. No pressure then. None at all.

"Are you worried about that?" he asked gently. I squeaked out an affirmation, since nodding would mess up the mold.

"Well, don't worry about that. We'll ask you to do some Kegel exercises and then just *think* about doing some Kegel exercises so we can compare the activations in the sensory cortex under the two conditions. Even if there isn't an orgasm, it will still be very useful to us."

"You'll do great," Wise concurred, as she slipped a bottle of CVS-brand lube into the pocket of her lab coat, a coat with many pockets. I imagined it held any number of items she could pull out to help me relax if needed—some chicken soup, a couple of Dramamine tablets, or perhaps even an fMRI-safe dildo, if that's what the situation called for. The thought made me laugh a little, but only on the inside. I did not want to have to remold my mask.

Getting to the Big O

A few hours later the party moved to the fMRI suite at the nearby University of Medicine and Dentistry of New Jersey Medical Center. I donned a hospital johnny and was pushed back into the scanner's tube, as ready as I would ever be to have an orgasm in an fMRI. Almost immediately I heard Komisaruk's voice through my headset, telling me to lie still so they could perform a ten-minute anatomical, which is a scan that obtains a precise set of "slice" images of my entire brain. These images would form the "screen" onto which they would project the active sites in my brain in order to precisely specify where the activity was occurring. The brain's relay stations and the pathways connecting them are as complex and exactly laid out as the buildings and roads in a city. So, just as a fire department can locate a fire in a specific building, brain imaging can locate a hot spot of activation in a specific brain structure.

"Just relax, Kayt," he said. "Go to sleep if you want."

The magnet started to spin around me. As promised, it was loud. Da dadadadadadadadadada. Click. Aunnnnnnnnnnnnnnnnnnnk. Clank. THUNK! THUNK! THUNK! The sound was not unlike a combination of jackhammers, a few amateur tap dancers, and a test of the emergency broadcast system at top volume all rolled into one. It lasted the majority of my session inside the scanner, which was approximately an hour and a half. Even with ear protection, I could feel each click, clank, and whir all the way down my spine. The situation was not conducive to sleep, but I tried to will myself into a relaxed state anyway. fMRI is all about staying as still as superhumanly possible, and relaxing does help.

When I first spoke with Wise weeks earlier, she gave me pre-scan homework. "Practice self-stimulating to orgasm as often as you can. And when you do, try to stay as still as possible," she instructed. "It's much harder than it sounds. Just the other week I had to hold one woman's legs down because she was thrashing around so much. I just prayed we'd get some usable data out of it." The magnet is unable to accurately track blood flow if there's too much movement; you would not know if a result was due to the stimulus (that is, the self-stimulation) or just extra noise from the movement. Given that an individual fMRI scan costs thousands of dollars, that noise-inducing movement is something researchers hope to avoid. And my at-home practice made it clear that staying as still as possible in this particular scenario wasn't going to be all that easy.

Just as I was starting to zone out, not into sleep exactly, but into something like it, the noises suddenly stopped. After the ten-minute cacophony, the lack of booming was almost as deafening. I was awake and paying attention now. A few moments later Komisaruk's smooth voice came through my ear phones again. "Now we're going to do thirty seconds of Kegel exercises, followed by thirty seconds of rest," he said. "And we'll do that a total of five times."

Kegel exercises, named for their creator and staunch promoter, Alfred Kegel, are simply contractions of the pelvic floor muscles, the same ones you use to start and stop the flow of urine. They are generally recommended for pregnant women and men with prostrate troubles. I know they were a favorite of my German obstetrician. She told me on my first prenatal visit that I should do my Kegels religiously, and concluded her instructions with the line, made all the more emphatic by her strong Teutonic accent, "Do your Kegels every day and keep incontinence at bay!" She didn't understand why her Ben Franklinesque rhyme cracked me up.

Once the clanking resumed, I started flexing my pelvic muscles.

Komisaruk wanted to compare whether actually doing Kegels versus just thinking about doing them activated the same brain areas. Some of his previous work suggested that the sensory cortex, at least where the genitals are concerned, could be activated by thought alone. If it held up, it would be a new and exciting result, and one that could potentially inform future treatments for those who cannot have orgasms, maybe even give researchers some new ways to study sexual desire.

My pelvic floor (and Kegel-fueled imagination) amply exercised, it was now time for the big show. Ready or not, I had to woman up and bring myself to orgasm. In a few minutes I would know if loud clanks and clicks, hospital johnnies, and a tight mesh head restraint could make the magic happen. I certainly hoped so. My type A personality meant I was pretty invested in succeeding. I did not want to be one of those few participants who were not able to do it.

"Okay, Kayt. We're going to take a three-minute rest and then get right to self-stimulating to orgasm," Komisaruk's disembodied voice came through loud and clear. "Can we get you some lube?"

I was going to need all the help I could get. Lube certainly couldn't hurt. "Yes, please," I replied. When I heard Wise enter the room, I reached my hand up and felt the lubricant, cool and slick, pool into the palm of my hand. I tried not to dribble any onto the floor before putting my hand back down beneath the blanket and hospital johnny to wait for the magnet to start back up again.

Da dadadadadadadadadada. Click. Aunnnnnnnnnnnnnnnnnnk. THUNK! I was instructed to spend the next three minutes relaxing and thinking of nothing. Instead I spent it stewing on my performance anxiety, giving myself a pep talk. I can give myself an orgasm under these circumstances. I think I can. I think I can. I think I'll come. I could do it, darn it. I would.

"Self-stimulate to orgasm starting now."

Hearing my cue, I took a deep breath and got to it. It may not have been romantic or sexy in there and, man, this mask thing was starting to get *really* uncomfortable, but I was going to orgasm no matter what. Focusing all my concentration, I powered through it, keeping my head as still as possible. A few minutes later I raised my hand to let Komisaruk know my orgasm had begun. I wouldn't say it was one of my best, but, hey, in my humble opinion, it still qualified. Sadly, the orgasm rated higher than a few I'd had during sex with one or two of my old boyfriends. I lowered my hand to signal my finish and, with it, let out a long breath of relief. If I could have reached around to pat myself on the back, heck, to pat myself anywhere except on my clitoris, I would have done so.

"Thanks, Kayt," Komisaruk said. "Please rest."

I did it. I now had a great story if anyone ever asked me to name the strangest place I've ever had an orgasm. And I had helped science while doing it. Triumph for all parties concerned!

After a few minutes Komisaruk's dulcet tones flowed across my headset once more. "Uh, Kayt, would it be possible to try to self-stimulate to orgasm again?" he asked tentatively.

"Again?" I asked with some confusion. Was there a glitch in the magnet? Did I do something wrong? Was my orgasm not orgasm-y enough for the O-Team?

"Well, the orgasm was a little short-lasting," he explained. "Would it be possible to do it again?"

Crap. All mental back-patting ended right there. I had managed to get 'er done, but my orgasm was not up to snuff. So much for my grand contribution to science. There was no way around it—I had to give it another go. "Sure thing. I can do it again," I said. This silly orgasm was becoming my albatross, my white whale. But I couldn't allow it to be. I was going to kick its butt this time around. Not only was I going to make it happen, I was going to find a way to *enjoy* it, damn it.

"Great, thanks," Komisaruk said. "Do you need some more lube?"

"No, thanks, I'm okay," I replied. Stopping and waiting for lube would only distract me. I needed to get this done *now*. You know, before I completely lost my nerve.

With the magnet whirring back to life with a clank and a bang, I took a deep breath and reached back down under the blankets. Short latency or not, my previous efforts made the process easier this time around. As strange as it may sound, I relaxed into my work and started to enjoy myself a bit. Maybe a little more than that. When I heard myself audibly moan, I hesitated, wondering if they could hear me in the control room. I shook off the notion and pushed on. By the time I heard myself moan again, I didn't care if anyone could hear me or not. I was in the moment—I was going to get there again. And this time I was going to get there with some style. I raised my hand to signal orgasm onset and then I rode the wave on home.

"Kayt, that was wonderful. Great length and latency," Komisaruk told me. "It was really, really great."

Feeling cheeky, I replied, "It wasn't bad for me either." I heard laughter erupt in the background. I tried not to think about what else they might have heard just a few moments earlier.

After being instructed to relax for the next ten minutes as the researchers finished up my scan, I closed my eyes and did exactly that. This time, despite the noise, I managed to catch a few Z's. It would seem an orgasm is an orgasm is an orgasm for me too.

It's All about Timing

Two months later, in sunny San Diego, I met Komisaruk and Wise at the Society for Neuroscience conference, which gathers approximately thirty thousand neuroscientists to discuss the newest advances in the field. Having created their orgasm time line, they presented the data from my time in the magnet, merged with data from eight other participants, in one of the conference's poster sessions. And they were doing so with a three-dimensional movie highlighting the time line of brain activation. Call it brain porn—many of the neuroscientists who stopped by to check out the film had no problem doing so. I doubt Komisaruk or Wise would either.

As I watched the film I was struck by the sheer amount of activation. There was a heck of a lot of it. I felt the same way I had when I viewed the composite image of my own brain during orgasm a few weeks earlier. A lot of areas were lit up, the warmest colors indicating the highest levels of activation. It was hard to decipher what it all meant. When I mentioned this observation to Wise, she laughed. "It's true, there is a lot of activation," she told me. "An orgasm really is a whole-brain kind of experience."

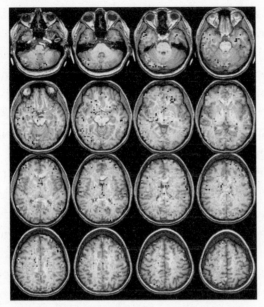

My brain at the point of orgasm. As Nan Wise said, it's a whole-brain kind of experience. *Photo adapted from data recorded by Barry Komisaruk, Rutgers University. Used with permission.*

No kidding. The group identified thirty discrete brain areas activated during self-stimulation to orgasm. *Thirty.* In women, at least, it seems we recruit a whole lot of our brain in order to get a little satisfaction. Given that extensive amount of activation, it can be a little complicated to assess exactly what all those brain areas are doing. However, Komisaruk was able to identify a few distinct areas that lit up before, during, and after orgasm.

What happened in my brain during orgasm? As I roughed up the suspect, so to speak, my genital sensory cortex, motor areas, hypothalamus, thalamus, and substantia nigra lit up. The hypothalamus was no surprise; it has consistently been implicated in all manner of reproductive behaviors, including arousal. The paraventricular nucleus, the part of the brain that produces oxytocin, is located there too. My motor areas controlled my fingers as I self-stimulated, and my genital sensory cortex registered that stimulation. My increasing heart rate could probably be attributed to that cerebellum activation. And the thalamus? It was integrating not only the activity of my wandering fingers but also the memories and fantasies I used to help build up my arousal. The substantia nigra, an area rich in dopamine-producing neurons, paired with the PVN's oxytocin release, had me feeling nice and relaxed.

Once I raised my hand to let the researchers know I was at the point of no return, my frontal cortex, that bastion of executive function, came online. Areas implicated in memory, integration of sensory information, and emotion also became active. As my orgasm came to a close, the hypothalamus turned back on, and reward areas like the nucleus accumbens and caudate nucleus were now flooded with dopamine. That was what gave me that final rush. The long and the short of it was that Komisaruk and his colleagues saw distinct temporal activity, with different brain areas being recruited as I went from arousal to orgasm and then back around again to rest.

"So did the results end up like you thought they would?" I asked Komisaruk, as a handful of other neuroscientists checked out the poster.

He smiled. "Some things were expected. The sensory regions become activated early on, and then the nucleus accumbens gets activated much later than the other areas. I expected that. But it was interesting: Some areas become active quite suddenly at the onset of orgasm, while others come on very gradually."

"Suddenly?" I asked.

"The inferior temporal gyrus is one area that comes on suddenly at onset," he said. "Same with the cerebellum."

Komisaruk went on to explain that the sudden cerebellum activation may have something to do with muscle tone. It is not a stretch—I tend to clench a bit at orgasm. Okay, maybe I clench a lot. But the inferior temporal gyrus? That is an area usually reserved for higher-level thinking and imagination, maybe fantasy. As far as I can remember, my eyes were shut tight, though I admit I was distracted and can't say for sure. Even if they were open, there are not a whole lot of compelling items to see inside an fMRI tube. But I was certainly imagining a few things as I achieved climax.

"There's some kind of cognitive process going on there, though we're not sure what it is," Komisaruk told me. "We saw a lot of cortical activation throughout the cortex which has not been reported in other studies."

One suggestion Komisaruk has for this cognitive process is a kind of inhibition. Perhaps those instructions to move as little as possible, any internal worries about my moans being overheard in the control room, all the little things I did to overcome my usual squirmy and loud orgasmic tendencies activated my frontal lobe. Or maybe, as I had no one else stimulating me to orgasm, that activation can be attributed to coordinating my own efforts. It is a puzzle yet to be solved.

Wise jumped in with her own revelations. "The process starts with very little activity and then gradually increases. That's not a surprise. But what was curious to me was that within the brain stem, we saw a lot of different patterns across participants. But there looked like there was a lot of similarity in the cortex. There is something going on there."

"We all have different kinds of touches that feel good to us," she continued. "In the future, we need more information about what people are really thinking about while they stimulate and perhaps a little more about what they're doing to get there."

An orgasm is an intense experience. Getting to that point involves a variety of cognitive, emotional, and sensory components—even when it's just you doing the work. Ultimately Komisaruk, Wise, and Whipple hope to understand enough about the different brain areas recruited to help individuals with anorgasmia, that unlucky percentage of women who never have orgasms. They openly admit, however, that we still have a long way to go before we understand the intricacies of it all.

"How do these different brain regions work together? How are they recruited differently in pleasure and pain? What can we learn from women who can think themselves into orgasm in order to help individuals who can't have an orgasm?" Wise speculated, "I can envision a time when people can regulate their own brain chemistry through some kind of internal process. But we're still in the infancy. Hell, we're still in the prenatal in this field. But I can't wait to see what will come in the next ten years. It's going to be amazing."

Though preliminary, the results astonished me. Orgasm is decidedly complex and affects multiple brain systems. I have no doubt researchers like Komisaruk and Wise will spend the rest of their careers trying to figure it out. Even if my postdivorce orgasm count remains on the low side, I can say I have done my part, no matter how meager, to help with the neuroscience.

Chapter 13

A Question of Orientation

The breaking headline as I journeyed home from Rutgers University after having my orgasm in the fMRI scanner was the tragic suicide of one of the university's own students, Tyler Clementi. His body had been recovered from the Hudson River that very morning, days after he had jumped to his death from the George Washington Bridge. His suicide was believed to be in response to his college roommate's posting live webcam footage of him and another male student having sex. This invasive act was apparently the culmination of several bouts of cruel bullying Tyler had faced from classmates because of his sexual orientation.

While I waited in the airline lounge in the hours before my flight home, local and national news shows highlighted the tragedy, along with the recent suicides of several other young gay men across the country. Most of these shows focused on bullying and the role it played in these terrible suicides, but the odd conservative pundit on one show used his airtime to insinuate that Clementi's death was nothing more than a bad (and perhaps expected) end to an immoral and unnatural lifestyle *choice*.

Joel Derfner is the author of *Swish: My Quest to Become the Gayest Person Ever and What Ended Up Happening Instead* and one of the stars of the Sundance Channel's reality television show *Girls Who Like Boys Who Like Boys*. When I told him I could not believe that anyone would toss out such views at this time of mourning, he sighed and said, "Sometimes I don't think they can help themselves."[1]

For decades the West has debated whether sexual orientation is something you are born with or a lifestyle choice. It is a sensitive and politically charged subject, especially since marriage equality has made its way onto the ballot in several states. Everyone seems to have an opinion about sexual orientation based on religious views, personal experience, or, as more results are published and then publicized, scientific research.

So what does neuroscience have to say about sexual orientation? So far, research suggests that it is determined even before birth. (Please note that bisexuality will not be discussed in this chapter. There has been very little neuroscientific research into this particular orientation, likely because there is such wide variability in behavior across those who identify as bisexual, making it very difficult to study.) "I haven't seen any evidence to suggest [homosexuality] is a choice at all," said Qazi Rahman, a researcher who specializes in the psychobiology of sexual orientation at Queen Mary University of London. "Yet the debate goes on."

To understand why debate continues, we need to take a closer look at the research—what little there is—that has been published. Historically the study of sexual orientation has focused on genes. Twin studies have long suggested that there is a genetic component to which gender a person ends up being attracted to; that is, genetic components are estimated to make up about 50 percent of the different possible variables that influence sexual orientation. However, as much as one might like to herald science as being completely objective, it is influenced by the culture and society of the day.

"People tend to think science is the truth, the end-all, be-all. It's objective: scientists have a hypothesis and then test it," said Steve Wiltgen. He is a postdoctoral fellow in the Department of Neurobiology and Behavior at the University of California, Irvine, and an openly gay man. When I met him he had recently presented a poster discussing the history of the neurobiological study of homosexuality at a neuroscience conference. "But the more you look back at the science that's been done, you see how the questions and the studies fit into the beliefs when that research took place."[2]

Historically, the neurobiological emphasis on so-called gay genes existed because homosexuality was considered a psychiatric disorder;

until 1973 it was even listed as one in the *Diagnostic and Statistical Manual of Mental Disorders*. Remnants of the belief that homosexuality is a type of disease still linger, if not in the scientific studies themselves, certainly in the way much of the general public chooses to interpret them. Clearly, early genetic studies looking at sexual orientation were colored by the idea that homosexuality was a biological mistake that should be undone. With this backdrop in place, some of the earliest work in neurobiology was based on the idea that if a gene or group of genes could be implicated in sexual orientation, there was the possibility of treatment, perhaps even a cure.

"Fruity" Genes

Chances are, if you have ever left your bananas on the kitchen counter for too long, you have seen *Drosophila melanogaster*, the common fruit fly, up close and personal. Fruit flies do appreciate overripe fruit. Though you may think only about ways to exterminate these annoying little insects, they are an animal model of choice for many neurobiologists. Their simple brain offers a good gauge to test a host of neurobiological theories, on topics ranging from learning to memory to sexual function. The history of the neurobiology of sexual orientation begins with this prototype.

These pesky fruit invaders are very, very heterosexual. Right-wing "family" advocates might do well to use these flies as a banner mascot next time they want to mount some sort of protest. Kulbir Gill, a geneticist working at Yale University in the early 1960s, discovered that by mutating a single *Drosophila* gene, he could turn these male bastions of heterosexuality into bisexuals who try to get busy with both females and males equally. In *Gay, Straight and the Reason Why: The Science of Sexual Orientation*, Simon LeVay, a scientist who studies sexual orientation, writes, "When these flies were put together in all-male groups they formed long moving chains resembling conga lines, with each male attempting (unsuccessfully) to mate with the male in front of it." It is quite a mental image. If you are a Carmen Miranda fan, it might almost allow you to believe scientists decided to nickname this gene "fruity" because of those dancing insects instead of any homophobia on their part. But not quite. Here is an example of how the stereotypes of the day can influence science. In any case, scientists would later refer to this gene as the slightly less loathsome "fruitless" (*FRU*).[3]

Although *FRU* was pronounced the first "gay" gene, it was a bit of a misnomer. Nearly thirty years later, when technology advanced to the point where geneticists could isolate and sequence *FRU*, they found that it did not determine sexual orientation per se. Rather it determined whether or not flies could discriminate between males and females. When scientists manipulated *FRU*, flies with the mutated gene tried to mount males and females indiscriminately. If you can't tell the difference between a boy and a girl, why wouldn't you try to mate with whichever fly happened across your path? That's pretty much what was happening in these mutated fruit flies. One would be hard-pressed to consider this finding a true analogue to homosexual behavior.

Since Gill identified *FRU*, several other "gay" genes have been identified in fruit flies and even in higher-order species, such as mice. Unsurprisingly, many of those genes are implicated in the processing of dopamine and glutamate, key neurotransmitters involved in love and sex behaviors. The discovery of those genes gave credence to the idea that scientists were on the right track—that a homosexual gene was out there, just waiting to be found. Over the next few decades, however, the background changed. With homosexuality removed from the list of disorders in the *DSM*, the scientific focus was no longer on a cure or treatment but on how sexual orientation may develop.

Recent work pinpointed a new possibility, a gene named "gender-blind" or *GB*, that expresses a protein that helps transport glutamate from neuron to neuron. In 2007 David Featherstone, a biologist at the University of Illinois at Chicago, mutated *GB* in fruit flies. Like *FRU*, this *GB* mutation changed the behavior of our stalwart hetero flies so that they tried to mate with both boy and girl flies indiscriminately. Again, the flies simply could not process the right sensory information to tell the difference between the two sexes. Featherstone and his colleagues hypothesized that changes to this gene might lead to changes in synaptic strength, and consequently produce an inability to process sexual stimuli.

Sure enough, when Featherstone used drugs or other genetic pathways to strengthen synapses, the flies were once again able to interpret the sensory stimuli and return to heterosexual mating behaviors. Many everyday Joes read about this study, along with the media brouhaha surrounding it, and deduced that sexual orientation was not biologically hardwired. Rather, it could be altered with drugs or gene therapies, if only scientists found the corresponding gene in humans. Whether or not that was Featherstone's intention wasn't the point. This study is still used by some as evidence that homosexuality is an alterable phenomenon.[4]

Derfner, who doesn't disagree when his friends describe him as "really, really gay," believes genes likely have something to say about how sexual orientation develops, just as they might for any other natural variation seen in behavior. But just because something might be alterable does not make it a choice. "I'm surprised that more people weren't up in arms about [Featherstone's] study," he told me. "Even if science develops to the point where we can change sexual orientation, and perhaps that is just a matter of time, I don't know, it doesn't address the question of whether we should."

When I asked him if, at any point in his life, he would have considered a change in sexual orientation had it been available, his answer was an immediate and emphatic "No."

One of the most recent "gay" gene discoveries occurred during a study of the regulation of sugar uptake, an enzyme called *fucose mutarotase*. Chankyu Park and his team of researchers at the Korea Advanced Institute of Science and Technology originally called this gene *FucU*, until a kind journal editor suggested *FucM* might be more appropriate. When the group knocked out *FucM* (or, as I like to call it, the gene formerly known as *FucU*) in mice, they found that it changed the sexual behaviors of females. The best part of this story is that the discovery was a complete surprise to the researchers. There was no original hypothesis that mating behaviors would be altered with the removal of the gene.[5] "It was an unexpected result, but it's very difficult to expect what the phenotype of a gene will be, even when you think you know its function," Park said to me. "We speculated that knocking it out might result in immune system issues, but even that was a guess."

Instead Park and his colleagues discovered peculiarities in mating behaviors in female knockouts. The male mice acted normally, getting busy whenever they were able. But the females, housed in cages with both male and other female animals, avoided the males at all costs. They did not assume a lordotic position during estrus and did not respond to or even sniff the male's urine. What's more, these knockout females tried to mount the other girls in a very male-like fashion.

"It's the first, to my knowledge, demonstration of a homosexual gene in females," Park said. He hypothesizes that *FucM* is influencing the proteins involved in sexually dimorphic brain development, with the result that those female knockouts have brain areas involved in sexual behavior that look more like the boys' sexually activated brain areas, particularly the preoptic area.

So scientists have found so-called gay genes, of several types, in fruit flies and mice. In some cases they have been able to reverse homosexual behaviors with drugs or the mutation of other related genes. A genetic correlate in the human genome, however, has not been found, and we cannot ignore the fact that the so-called homosexual behavior observed in these flies and mice is not strictly analogous to homosexual behavior in human beings.

It is a critical distinction. Gay human beings generally show a strict preference for one sex, their own, over the other. They can discriminate between the sexes; they just prefer their own. This occurs in both genders. Park's model may better mimic the behavior of homosexual women than previous genetic models, but the alterations had no effect on the males. Despite the hope that genes would shed some light on the differences between heterosexuals and homosexuals, there is no clear answer in the current research. The only conclusion that can be drawn from the current body of work is that there are some, as yet unknown underlying genetic correlates to homosexuality. And that may vary between males and females.

Looking to the Animal Kingdom

It might be easier to pinpoint which genes are involved in homosexual behavior (as opposed to finding them accidentally) if we could observe analogous behavior in other species. With so many overlapping behaviors between humans and our closest evolutionary relatives, primates, you would think that there'd be at least one monkey species that shows homosexual behavior. Although same-sex sexual behavior is observed widely in primate species—monkeys are happy to sexually cavort with members of the same sex—these animals never show a preferential choice for their own sex. And that preference for one sex over the other is what defines sexual orientation: gay men are attracted to other men, lesbian women are attracted to other women. Primates simply don't behave that way.[6] "This shows a difficulty in extrapolating animal models to humans," Kim Wallen said to me. "The requirement is that they consistently share features. When you look at same-sex behavior, the behavior you see in monkeys doesn't share a key hallmark with that you see in humans. The surface similarity doesn't fit with the substance of the phenomenon you are trying to study."

There is one exception in the animal kingdom: sheep. As it so happens, approximately 8 percent of rams, or male sheep, are same-sex-oriented. They still exhibit male-like mounting behaviors; they just happen to work that magic on other rams. That preference does not change even when these rams are castrated (though castration does slow their roll a little by reducing overall mounting attempts).[7] There don't appear to be lesbian sheep, however, so it is hard to draw exact parallels. A gay sheep gene has not been discovered, either—no analogue of the *FRU, FucM,* or *GB* gene has been demonstrated in this species.

Given the uniqueness of homosexual behavior in humans, many neuroscientists have opted to forgo animal models altogether and go straight to the source: humans. Alas, the work that has been done in human models has not yielded the desired results. "There is evidence that the X chromosome plays a role in sexual orientation," said Dick Swaab, a Dutch neuroscientist who has been leading the study of sexual orientation for decades. Genome-wide association studies of a genetic marker on the X chromosome called XQ28, by a lab at the National Institutes of Health led by Dean Hamer, suggest there may be a link to male homosexuality, but the findings have yet to be replicated.[8]

Despite this possible genetic marker, Swaab offered a word of caution: he doubts that there is a single gene at work in these behaviors and proposes that any genes associated with sexual orientation are most likely related to brain development processes occurring while an individual is still in the womb. Remember those his and her brains from chapter 6? Perhaps the development of those sexually dimorphic areas, while the fetus is still in utero, is altered or disrupted in some way, leading not only to different probabilities of disorders like schizophrenia and anorexia nervosa, but also to different sexual orientations.

Beyond the Bedroom

We have all seen the homosexual stereotypes. Gay men are too often portrayed as skinny, effeminate creatures who are into fashion and little dogs, while lesbian women get the thick and butch treatment. If you have gay friends or family members, you know these are oversimplified caricatures. Just as with heterosexuals, there is great variability in the gay community. Sure, you have your queens and your butches, but you also have folks you might not even know played for the other team if they hadn't told you. Despite this immense diversity, epidemiological and neuropsychological studies have pinpointed some interesting differences between heterosexual and homosexual populations.

On the cognitive front, homosexual men consistently demonstrate more difficulty with mental rotation and spatial perception tasks than heterosexual men; they are about on par with heterosexual women. They make up for this lack with better spatial location memory. Also like heterosexual women, they have good recall of spatial landmarks during navigation. They tend to do better on several language indices too. Rahman believes these cognitive differences suggest variation in the brain. "Mental rotation, for example, we know is dependent on the parietal lobe. Gay men perform on these tasks just like heterosexual women do," Rahman asserted. "What's interesting is that lesbian women don't differ on these tasks. That suggests the parietal areas may be organized differently."

Rahman argues that sexual orientation is not just about whom you are attracted to; it is a package deal, with a variety of other traits involved. These include differences in problem solving, spatial navigation, language, and social cognition. By testing those differences we may better understand the brain areas involved, as well as how these systems have evolved to grow together and complement each other.

This makes me think about Alexander Ophir and Steve Phelps's resident and wanderer prairie voles—their finding that the number of vasopressin receptors on certain spatial areas of the brain are more predictive of monogamy than the number of receptors on the reward circuitry. Reproductive behaviors may not involve just sex and bonding; spatial processing may somehow help the selection of successful reproductive strategies over time. That idea likely applies to homosexual behaviors too. It is certainly something to consider.

Neuroanatomical Differences

With neuroimaging techniques now more accessible to research labs, scientists can do more than just assume there are differences in the brain; they can test the hypothesis directly. The basic assumption in neuroscience is that all behaviors have an underlying neural correlate. If gene-based hypotheses about changes in normal, sexually dimorphic brain development hold any water, then researchers should see distinct differences between heterosexual and homosexual brains—and, as it turns out, they do.

Several structural and functional brain differences have been found between heterosexual and homosexual men, many of them in Swaab's lab. In 1992 Roger Gorski and Laura Allen at the University of California, Los Angeles, found that homosexual men have a larger anterior commissure, an area of the brain thought to be involved in language ability, than their heterosexual peers. Remarkably, the size of this area was more similar to that seen in heterosexual female brains. Ivanka Savic, pheromone researcher extraordinaire, noted sex-atypical asymmetries in homosexual brains, meaning that the homosexual brain looks a lot more like that of a heterosexual member of the opposite sex than like the brain of someone of the same gender. For example, straight men have asymmetric brains with larger right hemispheres; so do lesbians. The amygdala, the area responsible for emotional salience, shows similar connections in gay men and straight women, yet a different pattern of connectivity is shared by straight men and gay women. Other sexually dimorphic areas, such as the anterior hypothalamus, implicated in sexual behavior, and the suprachiasmatic nucleus, or biological clock, also show sex-atypical differences in homosexual individuals. Studies of sexual arousal suggest that separate brain circuits are activated in homosexual and heterosexual participants.[9] There is ample evidence that heterosexual and homosexual brains simply develop differently, resulting in dissimilar circuitry, cognitive skills, and behavior.[10] "We don't know if these differences are there because the connections in the brain are different or there is some effect of sexual practice or experience on brain organization," Savic told me. "But whatever it is, it's there and it's different."

Swaab hypothesizes that it all comes down to testosterone. Between six and twelve weeks along in a pregnancy, the Y chromosome, if present, starts a testosterone storm in the womb. The appearance of this extra testosterone results in the development of the penis and accompanying boy goodies. In its absence the female genitals will develop. Testosterone is not just for gonads. Along with estrogen and progesterone, it also has developmental and organizing effects in the brain. Sometime after the whole penis/vagina question is decided, these hormones, along with a cornucopia of other proteins and chemicals, influence the brain's development, resulting in sexually dimorphic areas—you know, those his and her brains. Puberty's accompanying rush of hormones seals the deal, activating those sex-specific brain circuits originally organized during this secondary process in the womb.

A glitch in that process (or perhaps even more than one) leads to differences in sexual orientation as well as sexual identity. Transsexuality, or transgenderedness (the term preferred by some in the lesbian, gay, bisexual, and transgendered community), is a condition in which a person with the outdoor plumbing of one sex strongly identifies with the opposite sex. These individuals often say that they were born in the wrong body, that their outsides do not reflect their true gender. It is an extreme condition—and one that can cause immense suffering for both the individual and his or her family. Like homosexuals, transgendered individuals also show sex-atypical brain differences, specifically in the hypothalamus and the bed nucleus of the stria terminalis, both linked to sexual behavior in rodents. In these areas transgendered individuals show patterns in volume and number of brain cells more like the gender they identify with than whatever gender they inhabit. This particular condition is still considered a disorder in the *DSM,* though many in the LGBT community believe it should not be. Swaab has maintained that although transsexuality should not carry a stigma, it is still a medical condition. Changing the body to match one's internal gender involves invasive sexual reassignment surgery, hormone therapy, and a heck of a lot of counseling. "You can live without problem as a homosexual—there is no reason to consider it a disorder," Swaab told me. "Transsexuality is different. It requires one to adapt to a body that can't adapt to the brain. That's a big problem."

What might account for brain development that results in homosexuality rather than transsexuality? The answer is unknown. These brain differences are not apparent until adulthood, so there is no way to detect them early in life. Scientists have suggested that mutations to genes that are responsible for a type of estrogen receptor, an androgen receptor, and aromatase, an enzyme that converts testosterone to estrogen, may all lead to abnormal hormone levels in utero, and, as such, are potential factors underlying transsexuality. However, any causal relationships between any of these and transsexuality cannot be determined. Something is happening in the brain's development; the making of that precise hormonal and chemical cocktail that results in matching gonads and brains is somehow off. What may be behind it all has yet to be determined.

Even leaving transsexuality out of the equation, there do not seem to be any easy answers to the development of homosexuality. It is unlikely that the glitch is the same in homosexual men and women. While studies suggest that excessive testosterone in the womb may lead to homosexuality in women, it is probably not working in isolation. Moreover, the situation in men is not quite as straightforward. Some studies suggest a lack of testosterone; others suggest that too much may lead to sexual orientation differences in men. Other studies indicate that testosterone isn't involved at all. Ray Blanchard, a professor of psychiatry at the University of Toronto, has proposed a different mechanism for homosexuality in men, one that involves an immune response in the womb.[11]

Want to know a fun fact about homosexual men? The estimated odds of being gay go up over 30 percent with each older brother you have. It's called the fraternal birth order (FBO) effect. Gay men generally have more older brothers than their straight contemporaries. It is a very robust finding and one limited to males. Having older sisters doesn't change your odds of being gay, nor is there any type of sibling effect seen in gay women. Here's another interesting tidbit: homosexual men with older brothers tend to have lower birth weight than heterosexuals. The combination of these findings led Blanchard to suggest that, over multiple male births, the maternal immune system starts producing antibodies against the fetus's antigens. This progressive immunization may result in variations in brain development and thus sexual orientation.[12] There is quite a bit of evidence to support this theory, but once again, it is difficult to prove either way. And since the FBO effect can account for only an estimated one in seven gay men, there may be a handful of different ways brain organization leading to homosexuality may be altered in the womb.

Smell You Later

What is it that attracts you to someone? I have met few people who are attracted only to a penis or a vagina, regardless of sexual orientation. They get there eventually, sure, but the attraction starts elsewhere. But where?

"The big question we still haven't answered is what sexual orientation is actually toward, what is it orienting to," Rahman said. "The intuitive response is gender or sex. But that could mean any number of things. It could be the shape of a body, a type of face, movement, something in the voice, or a combination of all these things. What's more, we don't know to what extent individual differences play a role in this orienting response."

There is a high probability that heterosexuals and homosexuals are simply processing sensory information—sights, smells, sounds, and tastes—a little bit differently. Take pheromones, for example. As I discussed in an earlier chapter, these bodily compounds are controversial, but they may tell us something about the different ways the heterosexual and homosexual brains process chemical messengers. Savic and her colleagues looked at brain activation using positron emission tomography in homosexual men, heterosexual men, and heterosexual women who had been exposed to androstadienone (AND), a compound found primarily in male sweat, an estrogen-like steroid (EST) that is found in female urine, and more common odors like lavender and cedar. Savic knew from previous studies that AND and EST activate the anterior hypothalamus in a gender-specific manner. What she did not know was if there would be differential effects based on sexual orientation.

It may not be much of a surprise, but that is exactly what she found. Homosexual men and heterosexual women showed hypothalamic activation in response to the AND, with maximum effect in the medial preoptic area and anterior hypothalamus. There was no significant difference in brain activation when the groups smelled the common odors.[13]

When Savic attempted the same study in lesbian women, she found the expected corollary. Lesbians and heterosexual men showed activation in the anterior hypothalamus in response to the EST; heterosexual women did not. Savic argues that this pair of studies provides strong evidence that sexual orientation shows differential processing of sex-related stimuli.[14]

What about transgendered individuals? Do they also show differences in their processing of hormones? Savic thought testing AND and EST in nonhomosexual male-to-female transsexuals was the next logical step. In case you are confused, that's a person in a male body who identifies as a female and has only ever had sex with women. The jargon can get tricky; intuitively you might think a male-to-female transsexual would have sex only with men. I did at first. But in this study Savic and her colleagues chose to define sexual orientation based on birth genitalia rather than any other factor.

When compared to heterosexual female controls, this group of male-to-female transsexuals showed very similar brain activation, but with an interesting twist. That is, the AND activated the hypothalamus in a similar cluster of neurons seen in women, while there was a small effect of EST activation in a different area of the hypothalamus similar to that seen in men. The latter effect was limited, but present. The data suggest, Savic argues, sex-atypical brain circuitry in the hypothalamus in these transgendered males.[15]

The same caveats apply in these studies as in any other pheromone study. Savic used a high level of pheromones, hundreds of times greater than you would find in actual sweat. Yet despite this, the differential patterns of activation observed in these different groups suggest there are important differences in the way the brain processes chemical messengers based on both sexual orientation and sexual identity.

Love Is Love

Much of the neuroimaging work concerning sexual orientation has focused on the ways the brains of homosexuals differ from those of heterosexuals in response to olfactory and visual stimuli. This is a good place to start because different things turn them on. But that's not to say there aren't some key similarities between the two groups. What happens if we focus just on feelings of love and ignore who happens to put what where during sex (which, frankly, probably differs as much within these groups as between them)? Remember Semir Zeki, the professor of neuroaesthetics at University College London who was one of the first researchers to try to map the neural correlates of love? He noted that hundreds of years of poetry and narratives describe love in similar ways. It did not escape his attention that those depictions remained similar regardless of whether the author was discussing love of the same sex or the opposite sex. Thus, he hypothesized, love should look the same across all brains, regardless of sexual orientation.

Zeki and his colleague John Paul Romaya scanned twenty-four people, twelve men and twelve women. All of the study participants admitted to being passionately in love, as well as in committed sexual relationships; half of both sexes just happened to be committed to members of the same sex. As in Zeki's first study, he and Romaya measured cerebral blood flow as individuals viewed photos of their partner and a familiar acquaintance of the same sex and approximate age.

Once again Zeki and Romaya found activation of the hypothalamus, ventral tegmental area, caudate nucleus, putamen, insula, hippocampus, and anterior cingulate cortex. And again there was extensive deactivation across the cortex. These results replicated Zeki's original study, as well as the idea that love is both rewarding and blind. The two could find no differences between the heterosexual and homosexual participants.

As far as the brain is concerned, passionate love is passionate love, no matter what gender you or your intended happen to be.[16]

What about Learned Behaviors?

With such a strong focus on epigenetics, I'm sure you are thinking there must be some kind of nurture component involved in the development of sexual orientation. It seems a logical guess. Perhaps certain learned behaviors or sexual experiences somehow further shape that neural circuitry laid down in the womb. After all, how often have you heard the old story that domineering mothers influence sexual orientation? But Swaab laughed at this notion. "When I give lectures at the medical school, I always ask the students who did *not* have a dominant mother to raise their hand. No one raises a finger," he said, continuing to chuckle. "I think having a dominant mother is far too prevalent to influence sexual orientation."

Just as there is no neurobiological evidence to suggest there is a choice when it comes to sexual orientation or sexual identity, there is nothing that indicates experience might have the ability to change it. "People always wanted to talk about choice. It's nonsense," Swaab claimed. "The truth is, the choice was made for you in the womb."

Based on his own personal experience, that of friends, and his research for his book, Derfner agrees that there was never a choice. He believes research into the genes involved in brain development sounds promising. "It makes sense to me, it feels true," he told me. "Obviously, though, I feel like it's probably all the factors people are talking about plus twenty-three more that haven't been discovered yet."

There is no single answer on sexual orientation or sexual identity, no clear-cut process in the womb that results in this behavioral variability. It's possible, if not probable, that a variety of different processes result in these behavioral phenotypes. No amount of domineering parents, show tunes, buzz cuts, or Harley Davidsons is likely to have the power to change them.

"It's not like every gay man wakes up one day and says, 'Gee, that Barbra Streisand is fantastic!' or 'I need to go see a musical and learn to knit,'" Derfner affirmed. "There is a lot of variation. And if we can leave politics aside, though I wonder if it's possible to do such a thing in America today, perhaps these variations can one day be explained biologically."

What's the Point?

At the neuroscience conference in San Diego, Steve Wiltgen told me he was asked one question, time and time again, while presenting his poster on the history of neurobiological research on homosexuality: "Should we be doing this kind of research?" He admitted he was hard-pressed to form a reply. "The truth is, I really don't know how to answer the question. When I started my own research, I wondered why this wasn't a more active line of study. There's so much to learn. Then I thought about the focus on cures and wondered what this information would tell us. If we did find a gay gene, what would society want to do with it? Since I can't answer that, I go back and forth about whether we should continue."

Like Wiltgen, Derfner is somewhat ambivalent yet hopes more knowledge may bring greater understanding. "The Ivory Tower geek in me says, 'Absolutely, study this.' But we're so far from knowing what's really involved. If we get closer in, say, one hundred or two hundred years, my answer might change. I might say no."

Wiltgen and Derfner should take heart in the fact that researchers like Swaab, Savic, and Rahman, unlike the early neurobiological pioneers in this field, have no interest in cures or treatments for homosexuality. Swaab contends the study of neuroanatomical differences between homosexual and heterosexual brains may teach us more about how all brains develop. "It's important to understand. By understanding, we can learn to accept it," he told me. "I believe acceptance of both homosexuality and transsexuality is very much improved because of the studies we've done here in the Netherlands. That means a lot to me."

Rahman agreed. He also suggested that the neurobiology of homosexuality may tell us something about the way brain organization has evolved over the ages. In fact he proposed that there could be a direct evolutionary benefit to having "gay" genes in the population. "People argue homosexuality is somehow thwarting evolution. That's not how it works. Evolution is about trade-offs between costs and benefits," he explained. "We know homosexuality comes with a package of other traits—perhaps having a number of gay alleles in a straight population carries some distinct advantages that can be passed along. If nothing else, we can assume gay genes are good ones because they are clearly successful. They've been reproduced in humans for a long, long time. And they will continue to be for a long, long time to come too."

Chapter 14

Stupid Is as Stupid Loves

You have heard it before: Love can make you stupid. It can make you take unnecessary risks. It can make you a little bit crazy. In fact we discussed it in the previous chapters of this book. Yet while love can certainly wreak havoc with behavior, it never acts alone. It's more of the team-player type.

The trouble seems to begin with attraction, that magnetic pull you feel toward someone long before any lasting bond has formed. Being drawn to another person, physically or otherwise, seems to have the power to cloud your decision making. Shakespeare said that love is blind, but perhaps sexual attraction is the thing that really limits our vision.

Studies that examine these kinds of effects often correlate these attraction-related deficits with hormones. Sexual attraction results in a rush of androgens and estrogens, those little chemical motivators (or, as Paul Micevych characterized them in chapter 5, gate openers). But does neuroscience support the idea that hormones make us stupid—in attraction or love?

The Little Head Thinking for the Big Head

Back in his stand-up days, Robin Williams drew uproarious applause for one line: "See, the problem is that God gives men a brain and a penis, and only enough blood to run one at a time." A plausible enough explanation for why men are often speechless in the presence of beauti-ful women, this joke seemed to find support when Dutch researchers from Radboud University published a study in mid-2009 demonstrating that attractive women could derail men's cognitive functioning. Not that most of us needed such proof. I'd wager many of us have seen the male attraction-associated stupidity in action at one time or another.

Johan C. Karremans, a member of the Department of Social and Clinical Psychology at Radboud University, starts the paper by describ-ing the study with a fun anecdote:

Some time ago, one of the male authors was chatting with a very attractive girl he had not met before. While he was anxious to make a good impression, when she asked him where he lived, he suddenly could not remember his street address. It seemed as if his impression management concerns had temporarily absorbed most of his cognitive resources.[1]

No mention of which of the two male authors, both accomplished men with upper-level degrees, happened to have this particular failing of memory. But the point is clear: anecdotally we all seem to know that beautiful women wield some sort of influence over men's reasoning, even those supersmart guys who should know better. That knowledge made Karremans wonder if there was something more to it than the funny thing that happened to that guy that one time back when he met a gorgeous woman.

To investigate this, Karremans recruited forty male students to com-plete what is called a two-back task. Here a stream of letters is presented on a computer screen, each for five hundred milliseconds, followed by a blank screen for two seconds. For each letter the study participant is asked to indicate, as quickly and accurately as possible, whether that let-ter is the same as the one presented two letters before it. If it matches, you press one key on the keyboard; if it does not match, you press another. A measurement of working memory, or the ability to actively keep infor-mation in your head so you can later manipulate it, the two-back is con-sidered to be a good indicator of basic cognitive ability.

After completing a baseline measure of the task, participants were led to an adjacent room, supposedly to pass some time as the next computer task was set up. Waiting there was either a male or a female "experimenter," a study confederate, or fake, there to engage the study participant in neutral conversation for a few minutes. After seven minutes of conversation, participants went back to the two-back task on the computer. Once finished, the participants were then asked to rate the attractiveness of the "experimenter" and indicate if they were currently involved in a romantic relationship.

It probably comes as no surprise that an attractive female "experimenter" led to a significant decline in cognitive performance; what's more, the more attractive the participant found the "experimenter," the worse he did on the task. Those results stood firm whether or not the participant was currently in a romantic relationship.

In a follow-up experiment Karremans and his colleagues tested both men and women on a different cognitive task and had them interact with one another instead of a study confederate. This time around, same-sex or mixed-sex pairs were thrown together and instructed to have a five-minute conversation. Afterward the two were tested on what is known as a "modified Simon task." This task presents words on a computer screen in the colors white, blue, or green. If the word appeared in white, participants were asked to determine whether it was a positive or a negative word by pressing a designated key. If it was in blue or green, participants were asked to ignore the word's meaning and just determine its color, blue or green. It may sound simple, but it is actually quite a demanding task, requiring one to switch quickly between two very different tasks. After participants completed this tricky business, they were given a questionnaire asking how much they might have wanted to impress their fellow study participant as well as whether they were currently in a romantic relationship.

Once again men, regardless of relationship status, did worse on this task if they had been chatted up by a hot girl beforehand. They also admitted to being much more interested in making a good impression when paired with a female. The more they wanted to impress the girl, it seems, the worse they did on the task. No effects were seen when they were paired up with another dude.

The kicker? Women did not show the same effects. Although those who admitted they were super keen on impressing their male cohort did perform slightly worse on the cognitive task, this result was nowhere near the correlation seen in the men. It did not take long for the results of this study to spread all over the Internet, with headlines like "Beautiful Girls Make Men Stupid"[2] and "Why Beautiful Women (Literally) Make Men Dumber."[3] Karremans and his colleagues argued that the study participants used up so many cognitive resources trying to impress the member of the opposite sex that there simply wasn't anything left for the task. (Note that the authors also suggested a similar study should be done with homosexual participants to see if the effect held when a gay man was interested in another man. It may not be the opposite sex that is interfering with cognition, after all, but the side effects of attraction itself.)

When the study first came out, I was still living with my ex-husband. When I mentioned the results to him in passing, he shook his head and said, "I'm pretty sure we all knew that already. I can't believe someone bothered to study this."

It is true that Karremans's findings do fit quite nicely with the common mind-set; it certainly comes off as something one might put in the *duh* category. But the results of two cognitive tasks don't exactly prove stupidity. It is possible that one might see an altered effect if a different cognitive measure was used.

There are plenty of distractions out there in the world that can get in the way of optimal cognition; one could make the argument that worries about a meeting at work or missing your lunch might also decrease performance on a cognitive task. Perhaps an attractive member of the opposite sex is not so special in that regard. Or perhaps there is something happening with hormone levels that interferes with the ability to do complex tasks.

"It's fairly easy to argue that men, in terms of the mating game, will have different strategies and goals than women, perhaps even that they are a little bit more immediate-minded," said Heather Rupp. At the Kinsey Institute she and Kim Wallen had demonstrated gender-specific brain activation patterns in individuals looking at erotic pictures. "But a lot depends on the kind of task you are asking people to do. Men and

208

women have different cognitive abilities. A raise in testosterone for a man after interacting with an attractive woman might be more detrimental to verbal-type tasks but might not affect women. But if you had a spatial task of some sort, you might see the opposite effect."

Yet Karremans and company did not measure testosterone in these males; they did not look at penile tumescence or even ask about arousal levels. Instead their conclusions were simple: If you have only so many cognitive resources at your disposal, and you use up a bunch of them trying to impress a pretty girl, you won't do as well on cognitive tasks. Fairly straightforward, really. But frankly, as much as we want to embrace the idea that a pretty girl has the power to make a man stupid, we still cannot answer a lot of the why's and how's here, including whether these results hold true across a whole battery of different cognitive tasks, across heterosexual and homosexual groups, or if they are somehow facilitated by our hormones. So although the headline "Beautiful Girls Make Men Stupid" is eminently quotable, it is not, when you look more closely at the data, exactly accurate.

Do Good Girls Like Bad Boys?

If beautiful women make men stupid, then the corollary is obvious: good girls like bad boys. Again, it feels true and is a phenomenon that, when studied, has the power to generate some seriously cutesy headlines.

As it so happens, there has been some work looking at whether women are more attracted to "masculine" men. When these studies are publicized the phrase *masculine man* often magically transforms into *bad boy*. This seems logical because these masculine men are usually rule breakers, a little rambunctious, and unimpressed by authority. They are charismatic and outgoing. They are often a little self-absorbed too. They can be the heroic type, pulling small children out of burning buildings or running across an active battlefield to save a fallen comrade. They are usually very healthy. The ladies seem to love them. And naturally they are chock-full of testosterone.[4]

Turns out we don't need to look over a man's résumé or even spend all that much time with him to know whether he's a masculine guy. We can get a pretty good idea from just taking a good look at his face. Those same testosterone levels correlated with those high-spirited behaviors listed above also have the power to carve out a particular kind of face. "It's all about the browridge and the jawline," said Rupp, before going on to tell me that the Australian actor Hugh Jackman is a good example of a masculinized face. "Another good one is the actor Javier Bardem. He's not even all that good-looking, but he's still somehow really hot."

I nodded my head in agreement. Javier Bardem is a great example of what screenwriters Heather Juergensen and Jennifer Westfeldt so poignantly defined as "sexy ugly" in the provocative 1990s film, *Kissing Jessica Stein*—that is, smoking hot without being traditionally cute. I certainly wouldn't throw him out of bed for eating crackers. "But with Javier Bardem, it's not just his face," I said. "He's got that accent, that deep, gravelly voice. And he carries himself with a lot of grace and confidence."

"That's a great point," she responded. "But those are also testosterone-related traits. We're talking about faces, sure, but in the real world, these testosterone-related traits go together."

Still, faces offer us a lot of information. Time and time again, neuroscientific studies have shown that the brain loves some faces. People are very good at both recognizing and distinguishing faces, even when they are incredibly alike. And it appears that a specific area of the cortex called the fusiform gyrus helps make sense of them. It's clear, faces are special.

Several studies have shown that women are more attracted to a masculine face when they are at their most fertile, a period of the menstrual cycle called the follicular phase. The idea here is that female hormones, particularly the raised estrogens, are suggesting that a girl go for a particular mate type, the masculine type. After all, those masculine genes with improved health and extra chutzpah would be beneficial to any potential offspring. If she is not ovulating, well, then a more feminized guy will do, one who is cute but not as strong in the brow and jaw. From an evolutionary perspective, that more feminized guy, without all that raging testosterone, would make a better helpmate; he'd be someone who would stick around, help raise the kids, and do his part in keeping the house tidy. It is a great story, but it is hard to directly tie these things together. The evidence suggests there is some kind of relationship between hormones and the kind of man a woman is attracted to, but how that effect is mediated is still up in the air.

To look at the underlying neural correlates of this phenomenon, Rupp and her colleagues, including Thomas James, a neuroscientist at Indiana University who studies perception, measured the brain activity in sixteen single female participants as they evaluated photos of male faces that had been slightly masculinized and feminized using computer morphing software. "When we put the two faces side by side, women didn't consciously see them as masculinized or feminized," James said. "It's very difficult to see the difference between the two."

The task was fairly simple. Each face was presented with the number of the man's past sexual partners and typical condom use. Study participants were asked to indicate whether they would be likely to have sex with the man based on the information given. Brain activity was measured both while participants were in the follicular phase of their cycle and then again in their luteal phase (the nonfertile half of the menstrual cycle). Their blood was also drawn at both sessions to get a more exact measure of their hormonal state.[5]

The group found a few interesting results. First, compared with the feminized faces, masculinized faces led to more activity in five specific brain areas: the left superior temporal gyrus, bilateral precentral gyrus, right posterior cingulate cortex, bilateral inferior parietal lobule, and bilateral anterior cingulate cortex. These areas have been implicated in face processing as well as the assessment of risk, suggesting that, consciously or not, masculinized faces are perceived as not only more attractive but also more dangerous. The effect was quite robust considering just how slightly the faces had been morphed.

When the group looked at hormone levels, they found that a woman's own level of testosterone positively predicted activation in the anterior and posterior cingulate, brain regions implicated in decision making. Yet which phase of the menstrual cycle a participant happened to be in was not correlated with any of the significant brain activations. The authors hypothesized that they had not been specific enough in the timing of testing. With a short window of fertility, they may have just missed the mark in catching the follicular phase. Still the findings were striking.

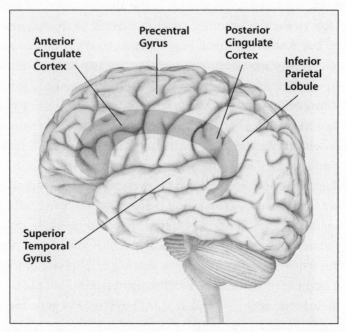

These areas were activated in the processing of masculinized faces.
Activation in both the anterior and posterior cingulate was correlated
with hormonal state. *Illustration by Dorling Kindersley.*

Those masculinity cues were fairly subconscious. "Both women and men are good at picking up nonverbal, subconscious cues about a potential sexual partner that tell them all kinds of things they aren't aware they are receiving," Rupp told me. "Your hormone level might make them more salient." But like her former advisor, Kim Wallen, Rupp was quick to point out that seeing a masculine dude during your follicular phase doesn't mean you are compelled to follow through. "You get these subconscious cues, yes, that may make you more likely to go home with a particular guy," she said. "But that's your subconscious. It doesn't have the only say. The conscious part of your brain can then come on board and help decide whether or not you are actually going to do it."

Rupp, James, and their colleagues wondered if perhaps hormone levels were changing the way the women appraised male faces. Perhaps the higher estrogen levels seen in the follicular phase increased motivation to see those masculinized faces, hot but risky, in a more positive light. They decided to test this with a second group of participants who would do the same task as in the previous study, but this time around the women also viewed photos of houses as a control condition.

In this follow-up study the researchers did see altered cerebral blood flow depending on a woman's cycle phase. When women were most fertile, in the follicular phase, they showed increased activation in the orbitofrontal cortex, an area of the brain implicated in reward and risk evaluation as well as judgments of attractiveness. This was observed when the women viewed both masculinized and feminized faces, leading Rupp to suggest that the follicular phase brings with it not only a lot more estrogen but also increased positive appraisal, or a higher likelihood of wearing rose-colored glasses when a girl interacts with a good-looking dude.[6]

Like Karremans's study, these results generated some great headlines, including "Explore Your Dark Side to Win Her Over"[7] and my personal favorite, "Why Nice Guys Finish Last."[8] Of course neither topic was really what Rupp, James, and their colleagues were investigating. So when I asked James, "Are girls more attracted to bad boys?" he laughed.

"My answer is that we don't know enough," he replied. "I know people want answers, usually a black-and-white answer. But science doesn't provide those."

Rupp gave me a similar response. "The verdict is still out. Obviously, for physical reproduction to occur we need these sweeping hormonal changes. And we also know these things affect perception and motivation. But we're still working with a lot of assumptions here. There are a lot of holes in our understanding that we need to fill."

Sex and Decision Making

Maybe some of those holes involve decision making itself. That is, are the decisions we make about sex fundamentally different from those we make about what to eat for breakfast or whether to invest in a particular stock? Despite the fact that the brain lights up like crazy for sexual stimuli, Rupp and James argue that these decision-making processes are not different from any other type.

When the group looked more closely at the anterior cingulate cortex activation during viewing of all male faces, they noticed something interesting. Study participants reported that they paid attention to how many sexual partners a man had and whether he was likely to use condoms; the lower the risk, the more likely the participants would be interested in having sex with the man. Those low-risk men elicited stronger activation in the anterior cingulate cortex, as well as in the midbrain and intraparietal sulcus, which are areas that light up in all manner of decision making, whether the task involves an economic game or whether it involves having to weigh risks and rewards. This, the researchers hypothesized, means that sexual decision making is no different from any other kind.[9]

"People tend to think of sexual behavior and sexual decision making as something unique that we don't understand," Rupp said. "The data, however, suggest otherwise. That makes sense—the brain doesn't do redundancy. And it would seem sexual motivation and decision making is not unlike any other type of reward-based decision making we do in life."

Rupp, James, and their colleagues set out to test this directly in a new study. I volunteered to be a pilot subject. Though a bit older than the age range they focused on, I could still assist them. I would do all the things a normal participant would and help them work out any kinks in the study paradigm. I traveled to Indiana University, as part of my pilgrimage to the famous Kinsey Institute, to have my brain scanned during my nonfertile luteal phase—this time to look at my decisions as opposed to my orgasms.

In this study I was shown four different kinds of photos: foods, alcoholic beverages, men, and everyday objects. Each photo, no matter what type, had a number score and the word *Yes* or *No* next to it. For any given food, the number represented the number of calories and the Yes/No told me whether the restaurant preparing the food had been previously cited for health violations. The values next to the drinks indicated the number of alcohol units and whether I had a designated driver on hand to take me home if I overindulged. Those next to the boys? The number of previous sexual partners and whether the guy typically uses condoms in his liaisons. And for the everyday objects, the number was a price and the Yes/No denoted whether the store that sells a product allows returns. My job was to evaluate the photos and their corresponding values and decide whether I would eat, drink, screw, or buy, respectively—just make a gut decision based on the information given. And I would be doing so using a 4-point scale—extremely unlikely, somewhat unlikely, somewhat likely, or extremely likely—tapping my responses into a keypad that would come with me into the scanner. It was a simple task, yet it took a bit of mental coordination to complete in the few seconds each photo was displayed.

Because I wouldn't be moving parts of my body to try to elicit orgasm, the fMRI setup was much simpler. To start, it was located not in a hospital but in the psychology building itself, a few doors down from the faculty offices. Very easy access, and no mesh S&M-type mask was required this time. My head was centered and stilled with some soft foam. The noise level was the same—there was no changing the sound of a spinning magnet. But I was infinitely more comfortable than I was in the orgasm study, which was a good thing considering I was going to have to think more. Not to mention press four different buttons, as opposed to just the one located on my nether regions.

Moments after the magnet started to spin, a small plus sign on the monitor let me know where I should be looking. Clank. Click. THUNK! THUNK! THUNK! I was ready. The first photo was of a fruity cocktail, some frozen strawberry concoction in a martini glass, garnished with a sparkling sugared pineapple. Next to it were the number 4 and the word *Yes*. Okay, this drink would be four alcohol units. That seemed like a lot (especially at eight in the morning), but I had a designated driver ready to give me a ride home. Still, I was not really in the mood for an ice cream headache. I clicked "somewhat unlikely" and waited for the next image to appear.

It was a photo of a young man. He was maybe twenty-two years old and straight out of an Abercrombie & Fitch catalog. His hair was too long for my taste, deliberately tousled and gelled to defy gravity. He needed a shave too. His corresponding values told me he had had only one sexual partner in the past month and used condoms. Obviously he was not a high-risk bed partner. Objectively he was a good-looking guy. But I couldn't get past the hair. Like some crotchety old lady, I wanted to tell him to get a haircut. The only thing I could do was click "extremely unlikely" and move on to the next photo.

Over the next thirty minutes I was shown images of brownies, Scotch tape, fried shrimp, cheesecake, various mixed drinks, clocks, macaroni and cheese, and a grocery store's worth of crap light beer brands. And yes, there were plenty of boys too. My use of the term *boys* here is intentional. These sweet young things, like the first one displayed, were all demonstrably handsome. I certainly cannot argue that point. Yet not one was my cup of tea. Between the piercing gazes (à la Zoolander's Blue Steel) and really stupid hair, I was consistently left cold. I found that no matter how safe they may be as sexual partners, I was wearing out the "extremely unlikely" button each time one appeared. I was beginning to feel like some weird old lady. Some weird old *asexual* lady. This is the age of the cougar, for goodness' sake—I should have been all over the young dudes. So when a cute boy with a somewhat normal expression on his face popped up, I said that I would be "extremely likely" to go home with him on principle. In truth, if we were in some club or café and he actually did approach me, I'd be more likely to hire him to babysit my kid than to head back to his place to hit the sheets.

During the entire session, aside from that one boy, I clicked the "extremely likely" button a total of five times. That's out of what must have been close to one hundred photos of various foods, drinks, boys, and objects. I figure I was either very discerning, steeped in my luteal phase, or simply ready for a nunnery thanks to age and my divorce. What was I willing to say I was "extremely likely" to enjoy that fine morning? Some pink Post-it notes (probably because I had just run out at home), a nice-looking rare steak, some absolutely decadent-seeming macaroni and cheese, and a carafe of red wine. Out of all of these photos, including the photo of the one acceptable boy, I have a very strong feeling my brain showed the strongest activation for the steak.

Later I talked to Rupp, the researcher who created the stimulus set for the study. When I mentioned that the boy photos did not really do it for me, she laughed. "I remember when I was choosing those pictures I felt like I was some old lady picking out boy porn," she said. "These guys are definitely meant for the younger women. I mean, at the very least they won't look like boys to them."

Later I met up with Julia Heiman, director of the Kinsey Institute and a coauthor with Rupp and James on these studies, and she asked what I thought of the stimuli. As I had been honest with everyone else about it, I brought up the stupid hair again. But after having a few hours to think about my experience in the magnet, I felt obligated to try to justify my lack of interest. "I'm in the middle of a divorce right now," I told her. "I have a feeling my optimal rewarding stimulus is going to remain a steak and a glass of red wine until it's done."

She laughed. "That raises an important point. I don't think how you responded today is how you'll respond forever. It would be interesting to measure your response today and then again in a year. Who knows, after that much time, you may be in a new relationship or playing the field. You may even kind of like those guys with the stupid hair."

While Heiman was obviously trying to make me feel better, what she said made me think. Rupp was careful to say that we have a lot of assumptions about sexual decision making. Could it change over time? She and her colleagues have tested women between eighteen and twenty-three. What if the group looked at younger women? Older women? It's possible our tastes change with time—and our decisions too. It's also unknown whether those decisions are correlated with hormonal states, which can fluctuate with both age and context. It's clear that the brain is always changing, with every new thing we learn, with every relationship we have. But we know little about how that occurs, and how it may be mediated by hormone levels, across the life span. Perhaps those "bad boys," if we can even call them that, are interesting only when we are young.

It is also noteworthy that these kinds of studies have looked only at women's sexual decision making. What might we see if we studied men? Might their sexual decision making show different effects at different ages? Will they have issues with a woman's hair? Is the poor decision making mediated by attractive women seen only when men are young and virile?

The answer to all of these questions is that we just don't know. Despite those fun headlines, most of these studies offer up a lot more questions than answers. And as for my own lack of appreciation for the bad boys, I can only take Heiman's advice and suggest that you ask me about them again in a year. Another birthday or a sexy romp with a new fella might make me come to appreciate the stupid hair.

Chapter 15

There's a Thin Line
between Love and Hate

What do a new mother and a soldier have in common? It almost sounds like a joke, one with a variety of different punch lines, which, in all likelihood, would not be very funny. But as it turns out, both soldiers and mothers exhibit what some call a "tend and defend" response when it comes to protecting their own: their kids or their fellow soldiers. A recent study suggests that this response is linked to oxytocin. Work with animal models has long suggested that oxytocin not only promotes a bond; it can incite a certain type of aggression too.

"Oxytocin helps with the bond, but it doesn't only bring love," explained Kerstin Uvnäs-Moberg, the Swedish oxytocin expert and one of the co-organizers of the first neurobiology of love meeting. "New mothers are very defensive, extra protective of their child against their surroundings. So is a man in love. He can be jealous and aggressive. There is a shift in feeling that is extra pleasing and calming, yes, but there seems to be a shift also in how and what we may perceive as threatening when we have made that bond."

You can see just how aggressive a little love can make us by looking at newly bonded prairie voles. When researchers conduct partner preference tests with the prairie voles, in which a bonded animal is placed in a Plexiglas cage with its mate and a stranger of the opposite sex after a separation, it doesn't take long for the bonded animals to get aggressive with the interloper. A pair-bonded male will fight another male to the death for his mate. Females, though not quite as aggressive, are also pretty territorial.

"Female prairie voles will not share their mate with another unfamiliar female," Sue Carter, that pioneering oxytocin researcher from the University of Illinois at Chicago, said. "We discovered that if you put two unrelated females and one male together, the male would mate with both. However, after mating and by the time the babies were born, there was only one female left. One female died. We could not tell the cause of death, because in most cases there was no obvious fighting. However, we could predict which of the females would die because the one who would live was always sitting between her and the male, apparently guarding the male. We could predict who would survive with a lot of certainty because that female would continually reposition herself so the other female just couldn't get near the male. It was quite frightening to realize that prairie voles could be 'stressed to death,' presumably by being ostracized. We stopped doing these experiments once we realized that one of those females was always going to die if we left them in this kind of situation."

Ouch. Love, as they say, is fierce. Both in romantic and maternal bonds, along with the union come increased levels of stress and aggression. Carsten De Dreu, a researcher at the University of Amsterdam, wanted to see what role oxytocin might play in that enhanced aggression—and whether it might transcend a love connection and explain in-group affiliation and out-group aggression as you see in wars and other types of human conflict. More specifically, De Dreu hypothesized that oxytocin modulates what he calls "parochial altruism," or trust and self-sacrifice in order to benefit the in-group with aggression to defend against and attack the out-group.

Parochial Altruism

De Dreu and his colleagues recruited men to participate in a variant of the so-called Prisoner's Dilemma. In the classic version of this task, two suspects in a crime are offered options for a deal. They are instructed that if one gives up his partner and agrees to testify against him, he will go free while his accomplice receives the full sentence—say, ten years. If both betray each other, then they will both get reduced sentences of five years. If both manage to stay quiet, they will get a veritable slap on the wrist—only six months on a lesser charge. Each must decide whether to rat out the other suspect or stay quiet. If the suspects are thinking about the group and not just themselves (and trust that their partner will also do the same), the choice is easy: stay quiet. Doing so means both partners will receive the least punishment. Despite this, most individuals simply give up their partner. Fairly quickly too; in individual economic terms, doing so makes the most sense.

In De Dreu's study, participants were randomly assigned to one of two three-person groups. This variation on the original Prisoner's Dilemma involved money. Each individual was given ten euros. Participants could then decide to share that money within their own group or between the two groups. If the participant kept the money for himself, each euro was worth exactly that: one euro. If the participant gave a euro to the within-group pool, an extra fifty euro cents would be added to each in-group member—so basically the euro was now worth 1.50. The participants had another option: giving to the between-group pool. If they did that, they not only gave each of their own group an extra fifty euro cents but took away fifty euro cents from each member of the opposing group.

I know it sounds complicated, but by looking at where each participant put his money, the researchers could gauge in-group trust and out-group aggression. For example, contributing to the within-group pool yields the highest benefit to the whole group, representing that in-group love. Contributing to the between-group pool hit the out-group where it hurts, taking away their money while still offering some benefit to the in-group. If the participant decided to keep his money for himself, that suggested he did not identify with his group at all; he was only looking out for himself. Participants snorted some oxytocin or a placebo (they were unaware which they were getting) before sitting down to play the game.[1]

De Dreu and his colleagues found that oxytocin amplified in-group love. That is, participants were more likely to contribute to the within-group pool if they had received a sniff of the neuropeptide instead of the placebo before making their decisions. Oxytocin, however, did not change how they contributed to the between-group pool. It would seem oxytocin did not increase or decrease the desire to stick it to the out-group, just to contribute to the overall common good of the in-group.

Furthermore, when participants were asked, after they made their allotments, what they thought their fellow group members had done, the researchers found that those who had sniffed oxytocin estimated that the others gave amounts much higher than the estimates of those who hadn't received the neuropeptide. There was no difference in making judgments about the out-group. This, the researchers suggest, means the oxytocin also increased trust within the group.

Oxytocin administration did not result in any all-out hate or aggression toward the out-group. Would it help foster a defensive state, as seen in mothers and newly bonded animals? To test the idea, De Dreu ran a follow-up experiment. These participants performed the same task, but were also offered an option to cooperate with a member of the out-group to promote intergroup giving—something that would benefit everyone, not just one's own group. Individuals on oxytocin were much less likely to cooperate with a member of the other group, citing a need to protect the in-group. It would seem that oxytocin does not promote aggression to attack, but does promote aggression to defend a bond from a potential threat.

De Dreu's study tested only men. Makes you wonder if you would see the same thing in women, doesn't it? My gut says yes, and any woman who has ever been ostracized by a "mean girl" clique knows what I'm talking about. De Dreu does not have an answer. He claims in the study write-up that "violent intergroup conflict more often involves males rather than females," so the experiments pertain to the more relevant sex. I'm not so sure. Female prairie voles, it should be noted, are also more aggressive after a pair-bond. Perhaps different situations affect men and women in different ways. In any case I contend that more research needs to be done to see just how separate men and women might be in this respect.

It is clear that oxytocin plays a role in certain types of aggression as well as love and pair-bonds. De Dreu believes this work illustrates oxytocin's role in soldiers' ability to work cohesively as a unit against a common hated enemy. Can economic games really tell us about that thin line between love and hate in a more personal context? Though De Dreu's study illustrates oxytocin's reach (not to mention its subtlety) and how it might influence both cohesion and defensiveness in certain types of social interaction, it does not tell us much about the neurobiology of hate. It also does not include all of the other chemicals that may be mediating these effects.

"There is no behavior that is just a mechanism of oxytocin or vaso-pressin or whatever chemical you have alone," said Craig Ferris, the aggression researcher from Northeastern University. "The question I always ask is what else is released, what else might the oxytocin be inter-acting with here. And I venture to guess there are fifty different things released, if not more."

Neuropeptides are team players—they don't work alone. Ferris was quick to point out that all aggression is context-dependent, which is going to affect which chemicals are released and, by extension, what kind of brain activation occurs. It is not all about the oxytocin. It's just not that simple.

Neural Correlates of Hate

What about brain regions? If there is a thin line between love and hate, whether defined by aggression or a more subjective emotional state, shouldn't we see it in analogous brain activation? As a follow-up study to his work on love, Semir Zeki of University College London decided to use fMRI to look at the brain regions underlying hate. He predicted that there might be some similarity between these two states that could be reflected in neuroimaging results.

"Both love and hate are strong biological sentiments," he explained. "They are both motivating factors too, that may push people to do great things and sometimes push people to do very evil things. They both can be all-consuming. Hate usually has a negative connotation in people's minds, but it's actually a biological phenomenon that serves to keep people together and to achieve things against others. It is just as worthy of study as love."

When I asked Zeki if he thinks hate is a drive—as many researchers have asserted—he said, yes, it is possible that it falls into the same cat-egory. "I think it is a negative drive, but it's a drive that can help a person achieve things, some of which can be quite useful and others which can be quite harmful. I think you can be driven by hate to pursue lines that are to the detriment of others, but you can also be driven to pursue lines that are to the benefit of oneself and to others."

Love and hate of an individual can often be linked. Think about it: How easy is it to hate a person you once loved, especially if your relationship ended on a sour note? As much as we are taught from an early age that hate is bad, it almost seems easier at times to summon that kind of intense emotion for someone you held so dear. Could there be a connection? I asked Zeki that very question. "There is so much ambiguity in the relationship between these things," he said. "Beauty can often lead to desire, which can lead to love. That's an interesting topic from the point of view of neurobiology, understanding that progression. Similarly, how love may transform into hate is equally as interesting. Certainly [love and hate] are often linked."

Zeki and his colleague, John Paul Romaya, scanned seventeen healthy individuals as they viewed photos of someone for whom they expressed a strong hatred as well as three acquaintances for whom they had neutral feelings (matched for sex and basic appearance as much as possible). To measure hate-based feelings and actions, Zeki and Romaya created a new survey measure. In writing the questions, they focused on three elements they saw as necessary to the feeling of strong hate: a negation of intimacy, or the desire to be as far away as possible from the hate object; a feeling of passion, or a demonstrable anger or fear of the hate object; and a "devaluation" of the hate object based on expressions of contempt. The resulting hate score could range from 0 (no hate at all) to 72 (hating with an all-consuming, fiery passion). Sixteen of the seventeen study participants confessed to intense hatred for an ex-lover or a work colleague. Number seventeen was a little different—she saved her hate for a famous politician. Though one could easily argue that these types of hate are very different, they were grouped together, since, in all cases, that hate was directed at a single individual.

Once participants were in the magnet, they viewed each face individually, the hated face as well as the three neutral faces, for approximately sixteen seconds each. Each participant was instructed to click a button once the face disappeared; there were no additional instructions about what the participants should feel or imagine when viewing each photo. It was just a simple passive viewing task.

As expected, all faces activated the fusiform gyrus, a brain area that has long been implicated in the perceptual processing of faces. When Zeki and Romaya compared the face of the hated object with the other faces, they found additional activation in several unique areas, including the medial frontal gyrus; the premotor cortex, an area necessary for the preparation of motor planning and execution; and the frontal pole, a structure implicated in predicting how others may act or react. Zeki interpreted this activation as evidence that the neural pathways for hatred are distinct from those of love. He further hypothesized that these areas make up a network, one that is important in focusing attention on the hated object, predicting potential behaviors, and preparing to attack or defend oneself against that person.

Zeki and Romaya also found some overlap with their previous love studies. Both the insula and the putamen, two regions that have shown up in several romantic love neuroimaging studies, also lit up when participants looked at a person they hated.[2]

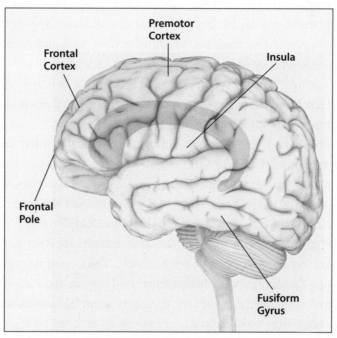

Brain areas activated during a hate task. The insula and putamen (not shown) are also activated in neuroimaging studies of love. *Illustration by Dorling Kindersley.*

I did not find this last result such a surprise. As my ex and I are sorting out the details of our divorce, there are times I am incredibly angry with him. I might even go so far as to say there are times I kind of hate his lousy guts. Arguing over money issues can do that to even the most saintlike person, which I decidedly am not, but the truth is, when I see his photo, particularly pictures in which he is smiling and laughing with my son, I feel something altogether different. I almost hate to admit it, but what I'm feeling is love. Whether I am simply remembering my past love for him or still carrying a bit of a torch, I cannot tell you. I'd imagine, whatever that feeling might be, it could have the power to confound the results if I were to participate in a study like Zeki's.

"Do you think perhaps you saw that putamen and insula activation because so many of the study participants were ex-lovers?" I asked him.

"I can't tell," he replied. "This was very much an initial study. I would like to carry on with these studies and compare people who hate someone because perhaps they disagree with them or have been harmed by them to those who hate someone because they loved them in the past."

"And would you say the results of this initial study support the idea, as the saying goes, that there's a thin line between love and hate?"

"I don't know," he said simply.

"No speculation?"

"No. I'm being very conservative about this. Now, there is obviously some connection. But what that might be, I don't know."

When I asked Zeki which finding in this study most surprised him, he was quick to reply. "The absence of deactivation in the cortex surprised me," he said. "Hate, like love, can lead people to irrational behaviors and actions. I was surprised we didn't see the same kind of cortex-wide deactivations as we did in the love study."

What might account for that lack of deactivation? Zeki is not sure, but it may have something to do with anxiety. The area of deactivation is close to one that has previously demonstrated involvement in obsessive-compulsive states. Perhaps the lack of deactivation has something to do with the obsession and compulsion necessary to keep hate alive. Let's face it, hating does require a bit of effort.

I couldn't help but also notice that the overlap between brain regions in love and hate fits in with animal studies showing the role of oxytocin in both pair-bonding and aggression. I asked Zeki if he thinks oxytocin is involved, if perhaps these brain areas are facilitated by oxytocin release. "I believe not only is oxytocin involved but also chemicals like dopamine, vasopressin, and serotonin," he said. "These are all strongly linked to one another, together in a fine balance. And that balance between the three seems to be very critical to determining states of love and hate."

To date Zeki and Romaya's study is the only one that has looked at the underlying neural correlates of hate. A unique study, it is one that, Zeki freely admits, calls for some follow-up work. He plans to continue looking at different kinds of hate, ranging from ex-lovers to racial divides, in future studies. As Ferris said, context is important—critical, really—to understanding what might be happening in the brain. A great love that turns into hate seems very different from a defensive posture against a hated out-group or a coworker who stole one's promotion.

Once again I nudged Zeki about the idea of a thin line between love and hate. Does he think the statement is true? Perhaps it is the backdrop of my divorce that prods me to keep pushing for an answer to this. Perhaps it's that I'm astonished to have a neuroscientist simply reply, "I don't know," without the benefit of some added speculation. Zeki chuckled when I asked again and then answered my question with a question. "How do you account for the ambiguity? That's another thing you must account for. There is a relationship between love, beauty, and desire. Hate too. Beauty often leads to desire, which can lead to love. Love may lead to hate. This transmutation is very interesting from the point of view of neurobiology."

"Definitely," I agreed. "Can we say anything about how that transmutation might occur, neurobiologically, that is?"

"That asks yet another question. Do we have adequate tools to study these questions?" he said. "I think we have some tools, but not all the tools we need. And science, good science, can only exist through having the proper tools to study particular questions. Right now, I don't know that we have the tools to address those questions. But, for now, we can say that hate, at least that which is directed at an individual, has a unique signature in the brain."

Chapter 16

———

The Greatest Love of All

A few weeks ago I picked up an airport rental car in Indianapolis on a brisk Sunday morning to make my way to the Kinsey Institute. I was too busy trying to find my way out of the multiterminal labyrinth to fiddle with the radio, and it was not until I found myself safely on the interstate that I realized it was set to a Christian station. Despite a preference for a little rock and roll when I am on the road, I listened for a few moments as a woman with a thick southern accent offered some religious testimony.

"My love for Jesus, like his for me, is all encompassing," she said. She then told the story of her spiritual rebirth, a tale of how she overcame the many bad decisions she had made before her salvation by letting go and allowing Jesus to lead the way. The details in her story aren't important, except that she credited her faith (and only her faith) for her life's turnaround. Her final words are what stuck with me; she ended her account with another proclamation of love. "There is nothing else on earth like opening your heart to a close and personal relationship with Jesus," she exclaimed. "It is true *ecstasy*."

I could not help but notice her choice of words throughout her testimony: love, all-encompassing, close and personal relationship. It was the last, ardent employment of the word *ecstasy* that really got to me. Change the context and this woman might well have been talking about a new boyfriend with a hot bedside manner, not her Lord and Savior. There seems to be a fair amount of overlap between the words we use to describe romantic love and those we employ in discussions of religious devotion. If we take a page from Semir Zeki's book, this correlation, present in religious writings and statements for the past few hundred years, may suggest comparable brain areas underlying these phenomena. Not for one faith, mind you; this is not just a Christian or Buddhist thing. It's that internal quest to understand something beyond ourselves and our world, however we happen to do it.

"Religion is very much related to culture," Mario Beauregard at the Université de Montréal told me. He is a researcher in the blossoming field of spiritual neuroscience, sometimes referred to as neurotheology. "But spirituality is different. The possibility of spirituality is in your genes, in your brain. By and large, it's something that is biologically possible."

Debates about the nature of the soul are outside the scope of this book. There are several other tomes out there that can take you through the long and somewhat strange neuroscientific history of the mind-body connection and how it may intersect with the soul. The focus here is more on whether religious devotion deserves the name "the greatest love of all" (as defined in my childhood Sunday school class, as opposed to the more popular Whitney Houston tune reference)—whether religious love shares some of the same neurobiological traits with some of the other forms of love we have discussed.

The "God Module"

Why do some people think religious, spiritual, or mystical experience has the power to make changes to the brain? Because everything does— these kinds of experiences are no exception. "There's nothing unique about religious experience in that respect," said Jordan Grafman, a neuroscientist who studies belief. "Any exposure to anything will change your brain a little bit. It's how you adapt and reorganize in accordance to new experiences that allows you to make it through life. No one should be surprised religion is any different."

Religion, like any other experience, has the power to change the brain. But what may be more interesting is that religious experience may involve specific brain areas. In 1997 Vilayanur Ramachandran, the director of the Brain and Perception Laboratory at the University of California, San Diego, made headlines when he presented evidence of a so-called "God module" in the brain during that year's Society for Neuroscience meeting.[1] Individuals with temporal lobe epilepsy, a condition marked by spontaneous and repeated seizures, often show intense religious devotion. Some have even proposed that religious prophets like Joan of Arc and Joseph Smith Jr. suffered from the condition, though those are posthumous diagnoses and therefore only conjecture. Regardless, there is a long history of epilepsy being paired with intense religious zeal.

Epileptic seizures are electrical in nature. They are bursts of overactivity in the brain that result in behaviors ranging from simple staring spells to violent convulsions. Ramachandran and his colleagues compared the brain activity of religious individuals with temporal lobe epilepsy, very religious individuals without the condition, and a nonreligious control group as they viewed words and images that were religious, violent, sexual, or neutral in nature. The measurement, a skin conductance response measure that indirectly indicates the strength of connections from the inferior temporal cortex to the amygdala, an area known to assign emotional meaning, might tell the researchers if the abnormal electrical activity in the epileptics' brains was "kindling," or strengthening neural pathways to add additional power to objects and words. That kindling effect might account for the fervor seen in these patients, as well as heightened brain response to certain types of words and pictures. If kindling was in effect, *everything*, from religious phenomena to athletic socks, would be more meaningful for these patients. Perhaps the epilepsy somehow upped that salience across the board.

Usually, in normal patients, the brain responds most to sexual stimuli. As Thomas James, the Indiana University professor running the appetitive decision-making study, told me, sexual pictures "get the brain going." Across the board the strength of response in the brain to sexual stimuli is usually two to three times greater than the response to any other type. If you take nothing else away from this book, at least you'll have learned that our brains are very, very interested in sex. In Ramachandran's study the same thing held true in the nonepileptic participants, regardless of whether or not they were religious: the sexual images and words jump-started their brains, or skin response, as it were.

In the epileptics, however, the pattern of activity was different. They showed heightened skin conductance response to religious words and icons, with diminished responses to the other kinds of stimuli—including items linked to sex. Ramachandran argued that the results demonstrated a localized area of the brain responsible for religious experience, most likely somewhere in the temporal lobe.[2] (For the record, Ramachandran also acknowledges the possibility that there is a God that visits these people directly. It is just a scientifically untestable theory.)[3]

At the same time Ramachandran was testing epileptic patients, Michael Persinger, a psychologist at Laurentian University in Canada, independently noticed that a group of neurons in the temporal lobe, near the amygdala, lit up when individuals were contemplating God or spirituality. It just so happened that this was the same general area in the brain Ramachandran was looking at in his epileptic patients. When Persinger stimulated these areas using a low magnetic current mimicking neural activity, something quite interesting happened: those folks reported feeling an overarching "presence" nearby, along with a heightened sense of well-being. Persinger argued that he had induced a religious experience using only a motorcycle helmet and some solenoids (a solenoid is a thin, coiled loop of wire that produces a magnetic field if treated with an electric current).[4]

Both these studies hit the press with force—Michael Persinger's so-called "God Helmet," especially. Many critics thought scientists like Persinger were attempting to reduce religious feeling to a simple neurobiological artifact. But most researchers in the field of spiritual neuroscience hope only to better understand the effect of religious experience on the brain. They are not interested in debunking or judging it.

"For me, the information Michael Persinger provided was that the temporal lobes play an important role in different types of religious and spiritual experiences," explained Andrew Newberg, director of research at the Myrna Brind Center for Integrative Medicine at Thomas Jefferson University Hospital and Medical College and a leader in this field. "I don't think the temporal lobes are the only mediators of those experiences. And

I don't think these results can tell us anything definitive about what the true nature of those experiences might be."

Although both Ramachandran's and Persinger's work suggests that connections from the temporal lobe to the amygdala are important in religious experience, neither methodology offers further details. Since the work was published, several neuroscientists have used various neuroimaging techniques to try to capture what is happening in the brain during an actual religious or spiritual experience. Mario Beauregard measured the brain activity of Carmelite nuns while they put themselves into a "state of union with God." These nuns live a cloistered life, filling their days with service and contemplative prayer. Beauregard calls them the "Olympic athletes of prayer"; each nun who participated in the study had cataloged thousands upon thousands of hours on her knees talking to the man upstairs. If that is not some kind of love, it's certainly an impressive commitment. For the neuroimaging study Beauregard and his colleagues isolated participants for approximately thirty minutes to allow them to relax into a mystical state, just as they might during a daily prayer session in the cloister, and then measured their brain activity.[5]

"These nuns believe that you cannot self-induce a deep, mystical state because it's the product of God's will. This is according to both their belief system and tradition," said Beauregard. "But they can enter into a moderate state of union if they are physically isolated. Their daily prayer practice, that daily experience, and their subjective reports of their time in the fMRI allow us to feel confident they were able to reach a meaningful religious or mystical state."

When the nuns were scanned while communing with their concept of God, several areas of their brain showed activation. Given that these experiences are complex and involve a variety of different types of imagery, it is not such a surprise. Like others in previous studies, Beauregard did see activation in the middle temporal area of the brain, which he and his collaborators argued might be related to the subjective experience of coming into contact with one's spirituality. Neuroimaging studies of Tibetan monks and other religious practitioners have also uncovered activation in this temporal area of the brain.

"When people are engaged in religious or spiritual practices, it affects many, if not all, parts of our body and brain," Newberg said. "Parts of the brain helping us with our ability to focus attention, to regulate our emotional responses and integrate social stimuli change during such practices. Changes occur in the body too that may impact how our autonomic nervous system works. And these experiences probably affect our hormone systems too."

There were a few other areas of interest that lit up in the brains of the Carmelite nuns in Beauregard's study, in particular the caudate nucleus, insula, and anterior cingulate. These are all brain regions that have been implicated in neuroimaging studies of romantic and maternal love.

Unconditional Love

When Beauregard talked to the Carmelite nuns after their time in the magnet, they all mentioned that they experienced a feeling of unconditional love while they prayed. Might that feeling, as they spiritually connected with their God, account for the brain activation? It made Beauregard wonder. Could there be a way to test the concept of unconditional love in individuals who had not devoted their life to God? As in other studies of love, the first challenge was to operationally define it. "Not everyone in the field believes in the concept of unconditional love, in this possibility," Beauregard said. Certainly when I queried friends about the concept, they were not so sure either.

Perhaps my friend Alyson summed it up best when she told me, "I believe in a lot of things about love. But unconditional love . . . I probably only believe in it when it's a parent's love for a child." Not being particularly religious but being head over heels in love with my kid, I tend to agree with her. But there are others who believe strongly in the concept of unconditional love outside the parental sphere. Sociologists and theologians who have studied the phenomenon find it to be paramount to the future of humanity. Stephen G. Post, president of the Institute for Research on Unlimited Love, postulates that, at its very essence, to experience unconditional love is "to emotionally affirm as well as to unselfishly delight in the well-being of others, and to volitionally engage in acts of care and service on their behalf without expecting anything in return." Those who are able to feel unconditional love do so freely and without expectation by definition, but they also admit to feeling very rewarded by acts done in its name.

One charity that attracts people who are able to love unconditionally is L'Arche. L'Arche creates homes of faith and friendship for disabled individuals and volunteer caregivers, called assistants, who commit to a year or more of sharing a home and life with an intellectually disabled charge. This organization not only believes in unconditional love but also makes the ability to feel and express such love one of the prerequisites for volunteering as an assistant. Beauregard recruited seventeen assistants from two local L'Arche communities near Montreal to participate in an fMRI study to identify the underlying neural correlates of unconditional love.

Each assistant was scanned as he or she looked at photos of disabled people. In one condition, participants were asked to just passively view the photos; in another, they were asked to generate a feeling of unconditional love toward the person depicted in the picture. Because of the rewarding aspects of unconditional love, Beauregard expected to see patterns of cerebral blood flow similar to those observed in earlier studies of romantic and maternal love.[6] "There is probably a common substrate in these various forms of love, especially with regard to the reward aspects of these different manifestations," he said. "But there are also some neural distinctions between them. It makes sense, considering there are differences experientially, the way people feel and experience these different types of love."

It should come as no surprise that this is precisely what Beauregard and his colleagues found. As in those early neuroimaging studies of love, the ventral tegmental area and the caudate nucleus, both significant parts of the brain's reward machinery, lit up like a Christmas tree. Maternal love's globus pallidus, another reward area, and the periaqueductal gray matter of the midbrain, a region with a large number of oxytocin receptors, were also active. Beauregard also saw significant activity in the anterior cingulate cortex, an area implicated in a variety of forms of love and attachment. Unconditional love, at least from a neural perspective, does seem deserving of the name. As we can see from the brain activity in these studies, there's quite a bit of overlap.

Beauregard also hypothesized that the dopamine system is involved.

"Some of the areas we saw activated are involved with the production of dopamine," he told me. "It still needs to be tested, but I would make the argument that there probably is an involvement of the dopamine neurotransmitter in this form of love, especially when you consider how rewarding the people tell us it feels."

Unconditional love also boasted some unique brain activation in comparison to other forms of love. Though the insula has been implicated in previous studies, Beauregard's work noted that a different part of it was active when participants generated unconditional love. He believes the activation was associated with reactions to viewing the photos. Previous studies suggest this part of the insula is necessary to integrate sensory and emotional representations as well as feel empathy. Several other areas mediating attention, visual processing, and a distinction between self and other were also activated.

"It's not well understood why this form of love is rewarding. It does not have the same characteristics as other kinds of love," said Beauregard. "But those who can feel it—and not everyone can—report it is a very rich, very rewarding experience and one that is a big part of their spiritual tradition."

Recall that Helen Fisher talked about love as a kaleidoscope—a different range of patterns in brain activation for sex, romantic love, and attachment. Perhaps unconditional love within a strong spiritual ritual or tradition provides a fourth system, with a familiar yet unique pattern. It is certainly possible—at least for those who can experience unconditional love. Yet what it is about a person's brain that offers this ability is unknown. As with most cases we've seen so far, further study is required.

A Brave New World of Love

In January 2009 Larry Young, the Emory University neuroscientist who studies pair-bond formation in prairie voles (and is arguably one of the most prolific researchers in this realm), published an essay entitled "Love: Neuroscience Reveals All" in the renowned journal *Nature*. In it he wrote:

> *The view of love as an emergent property of a cocktail of ancient neuropeptides and neurotransmitters raises important issues for society. For one thing, drugs that manipulate brain systems at whim to enhance or diminish our love for another may not be far away. . . . Perhaps genetic tests for the suitability of potential partners will one day become available, the results of which could accompany, and even override, our gut instincts in selecting the perfect partner. Either way, recent advances in the biology of pair bonding mean it won't be long before an unscrupulous suitor could slip a pharmaceutical "love potion" in our drink. And if they did, would we care? After all, love is insanity.*[1]

The essay generated a lot of talk, both in the neuroscience world and beyond. Some argued that Young had overreached, that even if the neurobiology had advanced to the point where we could create a so-called love potion (most agreed it had not), it was unethical, and perhaps even dangerous, to do so. Others, such as the *New York Times* writer John

Tierney, cheered in anticipation of "an anti-love potion, a vaccine preventing you from making an infatuated ass of yourself," that was supposedly on its way.[2]

When I visited Young and his lab full of loving prairie voles in Atlanta, I asked him about the essay. "Do you think a love drug or some kind of genetic test for love is something we actually could do at this point? More important, is it something we should do?"

"Should we? I don't think it's something we should go toward. No, I don't," he replied. "And it's not something we could do now, for sure. Given all this great work, the molecular and the behavioral and the genetic studies, we might be able to one day do these things." He paused for a moment. "But it would be a shame if we made a decision about another person based on a genetic test or some drug rather than our gut reaction. We could miss out on someone great."

In fact the neuroscientists I spoke with unanimously agreed that the creation of a Love Potion No. 9 based on the neurobiological study of love was a bad idea. Not only do we not yet know enough to create a drug that would work as intended, but, if we ever did get there, the risks involved in making changes to such an evolutionarily preserved brain system are immense. What about that love vaccine? You know, for those of us who want to avoid love and its myriad messy symptoms, like distraction, obsession, and even pain. Certainly the creation of a vaccine seems more benign than that of a love drug.

"That idea struck a chord with a lot of people. There are a lot of people walking around who just can't get someone off their mind," Young said. "I got a lot of letters about that one. In fact I got three handwritten letters from a man in Kenya. He read the *New York Times* article and wrote asking, 'Please, can you send me this vaccine for love? I need it.' He sent one letter every three months begging me for this vaccine."

I imagine the gentleman from Kenya is not the only one. Many of us who have felt love's keen sting may be just as desirous of a drug or a vaccine. If nothing else, I believe most of us are craving a few answers. Just a few biological clues, perhaps, about how we might better traverse love's muddy waters in a way that will allow us to open yet protect our delicate hearts.

If this were a self-help book, this is where I would tell you that I fell madly in love during the course of my research—that the things I learned as I explored the neurobiology of love, sex, and relationships helped me to finally track down my soul mate. Perhaps some hot neuroscientist who just knew we had matching oxytocin levels as soon as he strapped my head down for fMRI scanning. Or a successful doctor who just so happens to be my optimal major histocompatibility complex match as well as a guy with the right kind of *AVPR1A* variant for a loving and stable relationship. Perhaps I didn't even need to find someone new. Maybe my research into love helped me to better understand how my "dirty mind," or my unique epigenetic backdrop, played a role in the unraveling of my marriage. And with that understanding, my ex-husband and I are on our way to reconciliation, a neurobiologically based happily ever after. It would be a great clincher for a memoir or a cheesy made-for-TV movie.

Of course, none of that happened. That's just not how science—at least real science—works.

New research is revealing some truly remarkable stuff. We're learning a lot about how a variety of neurochemicals work together to make physical changes to our brains and about how our environment is involved in those changes. But the exact in's and out's remain an enigma, a puzzle yet to be solved.

"I think it's amazing that things that we never considered to be biologically based or chemically driven, like love, desire, and attachment, really are just that," Young told me. "There is a cascade of neurochemical events happening in the brain that cause us to feel the way we feel about another person and behave the way we do when we feel them. Sure, we have this cortex that allows us to think about things and to plan things—but underneath we have these ancient neurochemical systems that influence states we've long considered to be uniquely human. That's a big deal."

We're only beginning to understand that. Though there are some truly amazing findings in the literature, there aren't any that can be summed up in "Five Ways to Make Love Stay" or "Why His Brain Makes Him Cheat." Why not, you ask? The science published to date can offer you four very important reasons.

Because our brains are plastic. When I participated in the decision-making study at Indiana University, I did not find any of the young Abercrombie & Fitchesque boys all that attractive. At least, not attractive enough to want to say I'd go to bed with any of them. When I lamented about my lack of interest in those young, stupid-haired boys to Julia Heiman at the Kinsey Institute, she smiled and simply said, "I don't think how you are today is how you'll be forever."

She's right. Our brains are always changing. As our neurobiological insight into the brain grows, we're learning that it is incredibly plastic; that is, it is changing throughout our lives. With every new experience, every new item learned, every new relationship, there are subtle changes to our synapses. Over time those little changes add up to quite a lot. The brain I had when I first fell in love as a teenager is not the same one I have now. Neuroscientists are still working hard to suss out the details of that incredible plasticity, and they have a long way to go. But it is now clear that our brain is not a static thing.

"Plasticity is a funny thing," said Thomas James, the Indiana University professor who led the appetitive decision-making study. "It used to be thought that our brains were done growing around six years of age. That was it, they were done. And we're only starting to understand all the ways the brain remains plastic your entire life. That has huge implications for all of neuroscience research."

That includes the study of love. How I respond—how the chemicals in my brain respond—to a member of the opposite sex may very well change as I age. Not to mention after more extreme alterations to my brain and body that were brought on by pregnancy and childbirth. How it all changed, and how it will influence my neurochemistry and complex social behaviors both now and in the future, is just not well understood yet.

In the future, as more of the neuroscience community branches out and studies populations outside those easily recruited on the university campus, it's likely we'll see that this plasticity matters—and matters a lot—when it comes to complex social behaviors like love and monogamy.

Because our brains are complex. Neuroimaging studies have shown us that love has its own unique pattern of activation in the brain. But as Stephanie Ortigue, the neuroscientist from Syracuse University, reminded me, that pattern does not tell us the whole story. Neuroimaging measurement is currently limited by both speed and detail; as the technology evolves, we will learn more about how the neuroanatomical pieces of the love puzzle really fit together.

When I asked Craig Ferris, the aggression researcher who also happens to be an expert in neuroimaging, to describe the challenges of drawing conclusions from fMRI studies examining complex behaviors, he told me that current techniques lack the detail necessary to understand what the brain is really doing during any complex behavior. He used the amygdala, a brain area implicated in many love-related studies, as an example. "The rat has approximately twenty different subdivisions of the amygdala," he said. "People have spent lifetimes mapping just the amygdala in the rat, to understand the complexity of all these different subdivisions, what they do and what they project to. Each one of them does something different. It's a pretty complex little area. When you look at human neuroimaging work, however, you only see two parts of the amygdala mentioned: the left and the right. It's just a tough place to image in humans, so we're limited to that. But it begs the question, what does 'amygdala activation' really mean in these studies? Until we can get to a certain level of detail, to map out all the different parts of the amygdala in humans, I don't think we can really say that we know."

These measurements are a lot less explicit than we've been led to believe. Amygdala activation has been linked to understanding social cues as well as adding emotional valence to the outside world. It could, however, do much, much more. It likely plays a subtle role in a variety of other cognitive processes. As time moves on and with better technology, we will discover the extent of its function.

Beyond the tricky interpretation of neuroimaging studies, our understanding of the way different neurochemicals work when it comes to love is also limited by complexity. Oxytocin, dopamine, vasopressin, estrogen, testosterone—these are all chemicals that have been implicated in love and sexual behaviors. They are also the candidates for the love drugs and vaccines that many hope will soon be on the market. There's only one problem: they interact and cross-talk in ways that neuroscientists have yet to fully understand. They can lock up with one another's receptors. They can influence each other's production. And they are involved with many more bodily processes than just plain old love and sex.

When I asked Sue Carter, one of the pioneers in oxytocin and pair-bonding research, about the possibility of a love or a monogamy drug, she responded with some concern. "If we were to take what we already know about all the other biological systems, we'd immediately think twice about trying to fool Mother Nature with a drug," she explained. "Love is important. We need to be very careful. The chronic use of a drug meant to affect the oxytocin receptors, for example, has all kinds of interesting potentials that, as far as I know, have not been properly studied in any model. It is possible that there will be a down-regulation in the production of endogenous oxytocin. Over time the brain and the body might stop making the natural hormone. The second concern would be that the production of the oxytocin receptor might down-regulate. This would create a less reactive system. You would need more and more of the drug to get the same or maybe even lesser effects."

Until neuroscientists can better elucidate the how's and why's of the many ways these chemicals operate in the brain and the body, it's chancy to even consider taking a drug or vaccine. That includes snorting a little oxytocin spray, spraying yourself with some androstenone, or taking a so-called love-related brain chemistry supplement—all already available for sale on the Internet. "We just don't know what will happen if you interfere with the natural feedback loops, and so far this has not been properly studied even in an animal model," Carter said.

More important, simply upping one type of chemical is not going to give you the effect you want. It flies in the face of the fact that humans are much more than the sum of their neuropeptides. We need all of these chemicals—and a whole lot more—to experience love. "Biology is only a part of love. There is your culture, there is your background, and there's your huge cerebral cortex that can reason you out of it in the right circumstances," said Helen Fisher. "We are in the infancy of this field. We are only beginning to understand all the ways the brain may be involved with love. We are only beginning to understand just how complex it all is. It's going to take a lot of time to get to the bottom of it."

Because context matters. Our brains do not act in a vacuum. They are profoundly affected by our environment—right down to the neuron. As Moshe Szyf, the leading epigeneticist, said, "You can't just study the cell anymore. There isn't just a cell." All of our behaviors, including love and sex, can be understood only within the context of our environment. In fact at the level of our genes, those behaviors are actually regulated by our environment.

"The genome is set at a certain level. That's highly programmed, certainly," said Szyf. "But the early life environment sends signals to the genome saying, 'This is the kind of world this kid is going to live in. Let's program all these things to fit in with this world.' And then it resets the system. It makes very small changes in many, many different genes so the animal can adapt to the outside world."

And, as demonstrated by Szyf's and Michael Meaney's work, these include changes that have the power to alter our parenting and reproductive behavior. Blaming biology, as it were, for our relationship foibles does not cover it. Our environment and our relationships with others play an equally important role.

Because we are all individuals. This is the big one. Plasticity, complexity, and context all come together to make sure that each of us is unique in our own way. We're the product of our genes and our experiences. Don't think the sum of all that is not going to have some bearing on our love lives.

"The truth about sexuality, about love, about complex behavior in general, is that it is incredibly variable," said Julia Heiman. "So much so that many people say variability is the norm. And, you know, that fits in well with biology. Variability and biology go hand in hand."

In statistics a normal distribution of anything is represented by a bell curve. If scientists are talking about a "normal" type of love, a "normal" sex life, or a "normal" parenting style, it's never one value. Those behaviors are distributed across the curve, with the more "normal" values appearing at the curve's topmost point and the rest fanning out to the sides. "If we define normal as statistically normal, there's a wide range of what fits into the normal category," said Heiman. "There is also each end of a normal curve in which there are fewer cases—for example, people with very high or very low sexual desire. Are these at the less frequent extremes of the normal curve? Yes. Are they therefore bad or problematic? No, not necessarily."

We are all, each and every one of us, a little different. That is reflected in our epigenome, our neurochemicals, and our behavior. How could we not be individuals in all the ways we love as well? Those easy answers our hearts have been seeking from neuroscience—those rules, guidelines, and surefire methods to master love—simply may not be possible in the face of our own individuality.

The Great Mystery of Love

At a recent lecture series on sensory science and the arts at Johns Hopkins University, Semir Zeki, the first researcher to publish a neuroimaging study on love, commented, "The study of love is progressing quite well, at the molecular level and at the behavioral level." The key word is *progressing*. When I spoke to him on the phone a few weeks later, I asked him whether we'd ever solve the mystery of love.

"If we were ever to demystify love, we would simply replace the mystery with awe. Francis Crick said something to that effect many years ago," said Zeki. "Put yourself in the position of someone in the nineteen-twenties, interviewing a scientist and asking, 'Do you think we will ever demystify the mystery of life?' And then, in nineteen-fifty-three, Crick and Watson say, 'Well, all the secrets of life are contained in two strands of DNA with their base pairs.' That mystery was resolved in a way. And it was replaced by awe. I think, if we get there with love, it will be the same."

There's a line from Tom Robbins's cult classic novel *Still Life with Woodpecker* that has stayed with me over the years: "When the mystery of the connection goes, love goes. It's that simple."[3] If Robbins is right, we're safe: love isn't going anywhere. There's no danger, thus far, of unraveling its many mysteries, neuroscientifically or otherwise. I'd venture to say we also have plenty of time before we have to worry about dealing with any awe.

The neuroscientific study of love is moving forward, and it gains momentum each year. Yet even as new studies are published, they still lack the ability to answer our great questions about love. Think about it: Has DNA answered all your questions about the nature of life? I'm guessing not. It still retains more than a hint of the enigmatic. I believe love will be the same: no matter how far neuroscientists get in their study of this particular phenomenon, more than some mystery will remain.

"No one wants to hear it, but we know very little about how the brain functions in response to reward system experiences like sexuality and love. Not to mention that these things are not always rewarding," Heiman told me. "Readers need to be a bit more skeptical about the clever things said, the simple ideas about how the brain works and how it fits in with your everyday life. It's just not that simple. We need to trust the fact that things are probably a little more complicated—and to our benefit that they are complicated so we have the kind of flexibility and adaptability we do—and trust that they aren't complicated just because scientists want to make them that way."

Though we may wish it were so, there are simply no easy answers when it comes to love. There is no clever playbook for navigating love's messier situations; there are no promises to be revealed by five-step magazine stories or brain chemistry supplements. The brain is too complicated for that. That's the bad news. But take heart: that same neurobiological complexity is also the good news. It allows us to pass on the right information to our children about the kind of world they are likely to encounter. It lets our various neuropeptides stand in for one another when needed. It permits us the freedom to make choices, to adapt and change course with relative ease. It gives us the ability to take stock of any situation involving another human being, to remember our past experiences, to learn from them, and to calculate future risks. And it allows us to love, love, and love again—even after our heart has been broken. Really, that complexity is a blessing.

At the beginning of this book I promised no advice or guidelines regarding love. I won't go back on that commitment now. So if you still feel the need to pick up the next best-selling relationship book (there's always one or two making the rounds) or listen attentively to another episode of that syndicated advice show, I pass no judgment. I understand the need to hold on to something in the midst of all this crazy complexity. Perhaps, after reading these pages, you can take the guidance offered from these experts with an ounce of healthy skepticism, weed out the common sense from the hyperbole, and appreciate what studying the brain can really offer us when it comes to understanding love.

As for myself, I'll take the mystery. I don't mean to discount the neuroscience—not by any means. The research done to date has only whetted my appetite to learn more about the neurobiological basis of love. I fully intend to keep my eye on our friends the prairie voles, as well as the continuing neuroimaging work. Now that I find myself back in the big, bad dating world, I find the mystery to be a bit of a comfort. The lack of hard and fast answers means there is no "right" way to approach relationships, no one "right" partner for me. I have to admit that I find the notion quite liberating—a bit of love salvation, really. I am the singular product of my biology and my environment. Every potential mate can say the same. My brain's complexity offers me infinite possibilities when it comes to love. To my mind, that's much better than some precise how-to manual based on neurobiology, a one-size-fits-all list of musts and must-nots that has the potential to be as limiting as a straitjacket. There's already enough pressure involved in the search for love, thank you very much. I don't need the added burden of gene testing, neuropeptide measurements, or medications that may improve my love life at the expense of my cognitive function.

So, yes, I'll take all the mystery my neurons tender. At the very least, that's something I can count on.

Acknowledgments

—

I am greatly indebted to the researchers who allowed me not only to visit and explore their laboratories but also to pick their brains about the future neurobiological study of love. Many thanks to Sue Carter, Larry Young, Kim Wallen, Julia Heiman, Barry Komisaruk, Nan Wise, and Thomas James for their time and insights. I was also fortunate enough to interview dozens of other wonderful scientists over the course of my research, including Karen Bales, Katie Barrett, Mario Beauregard, Ray Blanchard, Lucy Brown, Joshua Buckholtz, Frances Champagne, Lique Coolen, Jeff Cooper, Andrea Di Sebastiano, Catherine Dulac, Craig Ferris, Helen Fisher, Michael Frank, Justin Garcia, Jill Goldstein, Ilanit Gordon, Jordan Grafman, Cynthia Graham, David Haig, Carla Harenski, Randy Jirtle, Pilyoung Kim, Sean Mackey, Hiroaki Matsunami, Bruce McEwen, Cindy Meston, Paul Micevych, Fernando Munoz, Andrew Newberg, Alexander Ophir, Stephanie Ortigue, Chankyu Park, Stephen Porges, George Preti, Qazi Rahman, Heather Rupp, Ivanka Savic, Wolfram Schultz, Charles Snowdon, Shannon Stephens, Dick Swaab, Moshe Szyf, Ei Terasawa, Kerstin Uvnäs-Moberg, Hasse Walum, Beverly Whipple, Shawn Wilson, Steve Wiltgen, Charles Wysocki, Jason Yee, Semir Zeki, and Marlene Zuk. I couldn't have asked for a more open and thoughtful group. Neuroscience truly is a field blessed with more than brilliance—everyone I spoke with on the course of this project was incredibly gracious and kind too.

Thank you to Demi Gandomkar, Denise Schipani, and Jen Miller for reading early drafts of my proposal, and to Carol Lee Streeter Kidd and her team for their transcriptional brilliance. Thanks to Kim Wallen, Todd Ahern, and Sara M. Freeman for some incredible images. Words cannot express my gratitude to Alyson English, Thomas Strickland, Nicky Penttila, and Joel Derfner for their comments as I wrote the manuscript.

They say every writer needs a good editor. They're right. I was lucky enough to have Sydney Tanigawa, Hilary Redmon, and Dominick Anfuso to help shape this into something worth reading. Every writer would also be lost without a good agent. Thanks to Joy Tutela of the David Black Agency and her assistant, Luke Thomas, for, well, everything.

I should also add that every writer needs good friends and family to cheer her on (and put drinks in her hand). I was more than well tended by Sarah Rose, Carl Morales, Dave Dillon, Jamie Pearson, Scott Collins, Aaron Bailey, Alison Buckholtz, Lily Burana, Rachel Weingarten, Kim Place-Gateau, Sylvia Hauser, Hillary Buckholtz, Shawn Gorrell, Clorinda Velez, Helen and Gayle King, the swinging Büdingen crew, Eleanor Jakes Willis, Tyler Schill, Reed Schill, and Max Schill. I love you all. I also appreciated all of the encouraging comments (as well as the answers to random personal questions about relationships) along the way from my friends, real and virtual, on both Twitter and Facebook—with a special shout-out to John Miller. Your kind and illuminating words helped more than you know.

Roan Low, I'd be lost without you. You're the best male girlfriend a girl could ever have.

Chet, you are an inspiration and one amazing kid. I promise to never again refer to you in print as "sexy." But most of all, thank you to my mother, Laurel Willis Sukel. I would not have gotten past the first page without her love, encouragement, and most excellent babysitting skills.

Notes

___

Chapter 1: The Neuroscience of Love: A History (Theirs and Mine)

1. McEwen, B. Meeting report: Is there a neurobiology of love? *Molecular Psychiatry*. 1997, 2(1): 15–16.

2. Nichol KE, Poon WW, Parachikova AI, Cribbs DH, Glabe CG, and Cotman CW. Exercise alters the immune profile in Tg2576 Alzheimer mice toward a response coincident with improved cognitive performance. *Journal of Neuroinflammation*. 2008, 5:13.

Chapter 2: The Ever-Loving Brain

1. Kolb B and Whishaw IQ. *Fundamentals of Human Neuropsychology, Sixth Edition*. 2008. Worth, New York.

2. Finger S. *Origins of Neuroscience: A History of Explorations into Brain Function*. 1994. Oxford University Press, New York.

3. Sizer N and Drayton HS. *Heads and Faces and How to Study Them: A Manual of Phrenology and Physiognomy for the People*. 1892. Fowler & Wells, New York.

4. Fisher HE, Aron A, Mashek D, Li H, and Brown LL. Defining the brain systems of lust, romantic attraction, and attachment. *Archives of Sexual Behavior*. 2002, 31(5): 413–19.

5. Bartels A and Zeki S. The neural basis of romantic love. *Neuroreport*. 2000, 11(17): 3829–34.

6. Aron A, Fisher H, Mashek DJ, Strong G, Li H, and Brown LL. Reward, motivation and emotion systems associated with early-stage intense romantic love. *Journal of Neurophysiology*. 2005, 94(1): 327–37.

7. Ortigue S, Bianchi-Demicheli F, Hamilton AF, and Grafton ST. The neural basis of love as a subliminal prime: An event-related functional magnetic resonance imaging study. *Journal of Cognitive Neuroscience*. 2007, 19(7): 1218–30.

8. Ortigue S, Bianchi-Demicheli F, Patel N, Frum C, and Lewis JW. Neuroimaging of love: fMRI meta-analysis evidence toward new perspectives in sexual medicine. *Journal of Sexual Medicine*. 2010, 7(11): 3541–52.

Chapter 3: The Chemicals between Us

1. Azevedo FA, Carvalho LR, Grinberg LT, Farfel JM, Ferretti RE, Leite RE, Jacob Filho W, Lent R, and Herculano-Houzel S. Equal numbers of neuronal and nonneuronal cells make the human brain an isometrically scaled-up primate brain. *Journal of Comparative Neurology*. 2009, 513 (5): 532–41.

2. Pani L and Gessa GL. Evolution of the dopaminergic system and its relationships with the psychopathology of pleasure. *International Journal of Clinical Pharmacological Research*. 1997, 17(2–3): 55–58.

3. Aragona BJ, Liu Y, Curtis JT, Stephan FK, and Wang Z. A critical role for nucleus accumbens dopamine in partner-preference formation in male prairie voles. *Journal of Neuroscience*. 2003, 23(8): 3483–90.

4. Curtis JT, Liu Y, Aragona BJ, and Wang Z. Dopamine and monogamy. *Brain Research*. 2006, 1126(1): 76–90.

5. Ferguson JN, Aldag JM, Insel TR, and Young LJ. Oxytocin in the medial amygdala is essential for social recognition in the mouse. *Journal of Neuroscience*. 2001, 21(20): 8278–85.

6. Uvnäs-Moberg K, Arn I, and Magnusson D. The psychobiology of emotion: The role of the oxytocinergic system. *International Journal of Behavioral Medicine*. 2005, 12(2): 59–65.

7. Carter CS. Neuroendocrine perspectives on social attachment and love. *Psychoneuroendocrinology*. 1998, 23(8): 779–818.

8. Esch T and Stefano GB. The neurobiology of love. *Neuroendocrinology Letters*. 2005, 26(3): 175–92.

9. Marazziti D, Akiskal HS, Rossi A, and Cassano GB. Alteration of the platelet serotonin transporter in romantic love. *Psychological Medicine*. 1999, 29:741–45.

10. Marazziti D and Canale D. Hormonal changes when falling in love. *Psychoneuroendocrinology*. 2004, 29(7): 931–36.

11. Emanuele E, Politi P, Bianchi M, Minoretti P, Bertona M, and Geroldi D. Raised plasma nerve growth factor levels associated with early-stage romantic love. *Psychoneuroendocrinology*. 2006, 31(3): 288–94.

12. Marazziti D, Del Debbio A, Roncaglia I, Bianchi C, Piccinni A, and Dell'Osso L. Neurotrophins and attachment. *Clinical Neuropsychiatry*. 2008, 5(2): 100–106.

13. Gray J. *Venus on Fire, Mars on Ice: Hormonal Balance—The Key to Life, Love and Energy*. 2010. Mind.

Chapter 4: Epigenetics (or It Is All My Mother's Fault)

1. Ebstein RP, Israel S, Chew SH, Zhong S, and Knafo A. Genetics of human social behavior. *Neuron*. 25 March 2010, 65(6): 831–44.

2. Gregg C, Zhang J, Butler JE, Haig D, and Dulac C. Sex-specific parent-of-origin allelic expression in the mouse brain. *Science*. 6 August 2010, 329(5992): 682–85.

3. Gregg C, Zhang J, Weissbourd B, Luo S, Schroth GP, Haig D, and Dulac C. High-resolution analysis of parent-of-origin allelic expression in the mouse brain. *Science*. 6 August 2010, 329(5992): 643–48.

4. Waterland RA and Jirtle RL. Transposable elements: Targets for early nutritional effects on epigenetic gene regulation. *Molecular and Cellular Biology*. 2003, 23(15): 5293–300.

5. Zhang T-Y and Meaney MJ. Epigenetics and the environmental regulation of the genome and its function. *Annual Review of Psychology*. 2010, 61:439–66.

6. Harlow NF. The nature of love. *American Psychologist*. 1958, 13:673–85.

7. Szyf M, Weaver IC, Champagne FA, Diorio J, and Meaney MJ. Maternal programming of steroid receptor expression and phenotype through DNA methylation in the rat. *Frontiers in Neuroendocrinology*. 2005, 26(3–4): 139–62.

8. Cameron NM, Shahrokh D, Del Corpo A, Dhir SK, Szyf M, Champagne FA, and Meaney MJ. Epigenetic programming of phenotypic variations in reproductive strategies in the rat through maternal care. *Journal of Neuroendocrinology*. 2008, 20(6): 795–801.

Chapter 5: Our Primates, Ourselves (or Why We Are Not Slaves to Our Hormones)

1. Ashby EA. *Puberty Survival Guide for Girls*. 2005. iUniverse, Lincoln, Neb.

2. Marazziti D and Canale D. Hormonal changes when falling in love. *Psychoneuroendocrinology*. 2004, 29(7): 931–36.

3. Guéguen N. Menstrual cycle phases and female receptivity to a courtship solicitation: An evaluation in a nightclub. *Evolution and Human Behavior*. 2009, 30(5): 351–55.

4. Miller G, Tyber JM, and Jordan BD. Ovulatory cycle effects on top earnings by lap dancers: Economic evidence for human estrus? *Evolution and Human Behavior*. 2007, 28:375–81.

5. Money J. Unpublished archive of John Money at the Kinsey Institute for Research in Sex, Gender, and Reproduction Library, Bloomington, Ind.

6. Micevych P and Dominguez R. Membrane estradiol signaling in the brain. *Frontiers in Neuroendocrinology*. 2009, 30(3): 315–27.

7. Garcia-Segura LM. Aromatase in the brain: Not just for reproduction anymore. *Journal of Neuroendocrinology*. 2008, 20(6): 705–12.

8. Kuo J, Hariri OR, and Micevych P. An interaction of oxytocin receptors with metabotropic glutamate receptors in hypothalamic astrocytes. *Journal of Neuroendocrinology*. 2009, 21(12): 1001–6.

Chapter 6: His and Her Brains

1. Eliot, L. *Pink Brain, Blue Brain: How Small Differences Grow into Troublesome Gaps—And What We Can Do about It*. 2010. Mariner, New York.

2. Fine C. *Delusions of Gender: How Our Minds, Society, and Neurosexism Create Difference*. 2010. Norton, New York.

3. Money J. Unpublished archive of John Money at the Kinsey Institute for Research in Sex, Gender, and Reproduction Library.

4. Cahill L. His brain, her brain. *Scientific American*. May 2005, 40-47.

5. Cahill L. Why sex matters for neuroscience. *Nature Review Neuroscience*. 2006, 7: 477–84.

6. Goldstein JM, Jerram M, Abbs B, Whitfield-Gabrieli S, and Makris N. Sex differences in stress response circuitry activation dependent on female hormonal cycle. *Journal of Neuroscience*. 2010, 30(2): 431–38.

7. Hamann S, Herman RA, Nolan CL, and Wallen K. Men and women differ in amygdala response to visual sexual stimuli. *Nature Neuroscience*. 2004, 7(4): 411–16.

8. Rupp HA and Wallen K. Sex differences in response to visual sexual stimuli: A review. *Archives of Sexual Behavior*. 2008, 37(2): 206–18.

9. Rupp HA and Wallen K. Sex-specific content preferences for visual sexual stimuli. *Archives of Sexual Behavior*. 2009, 38(3): 417–26.

10. McCall KM, Rellini AH, Seal BN, and Meston CM. Sex differences in memory for sexually-relevant information. *Archives of Sexual Behavior*. 2007, 36(4): 508–17.

11. Zeki S and Romaya JP. The brain reaction to viewing faces of opposite- and same-sex romantic partners. *PLoS ONE*. 2010, 5(12): e15802.

Chapter 7: The Neurobiology of Attraction

1. Zhou W and Chen D. Encoding human sexual chemosensory cues in the orbitofrontal and fusiform cortices. *Journal of Neuroscience*. 2008, 28(53): 14416–21.

2. Wyatt TD. Fifty years of pheromones. *Nature*. 2009, 457:262–63.

3. Grammer K, Fink B, and Neave N. Human pheromones and sexual attraction. *European Journal of Obstetrics and Gynecology and Reproductive Biology*. 2005, 118(2): 135–42.

4. Wysocki CJ and Preti G. Human pheromones: What's purported, what's supported. White paper prepared exclusively for the Sense of Smell Institute. July 2009.

5. Wedekind C, Seebeck T, Bettens F, and Paepke AJ. MHC-dependent mate preferences in humans. *Proceedings of the Royal Society of London, Biological Sciences*. 1995, 260(1359): 245–49.

6. Jacob S, McClintock MK, Zelano B, and Ober C. Paternally inherited HLA alleles are associated with women's choice of male odor. *Nature Genetics*. 2002, 30:175–79.

7. Keller A, Zhuang H, Chi Q, Vosshall LB, and Matsunami H. Genetic variation in a human odorant receptor alters odour perception. *Nature*. 2007, 449:468–72.

8. Savic I, Berglund H, Gulyas B, and Roland P. Smelling of odorous sex hormone-like compounds causes sex-differentiated hypothalamic activations in humans. *Neuron*. 2001, 31(4): 661–68.

9. Savic I, Hedén-Blomqvist E, and Berglund H. Pheromone signal transduction in humans: What can be learned from olfactory loss. *Human Brain Mapping*. 2009, 30(9): 3057–65.

10. Eastwick PW and Finkel EJ. Sex differences in mate preferences revisited: Do people know what they initially desire in a romantic partner? *Journal of Personality and Social Psychology*. 2008, 94(2): 245–64.

11. Springen K. The real laws of attraction. *Newsweek*, 14 February 2008.

12. Ireland ME, Slatcher RB, Eastwick PW, Scissors LE, Finkel EJ, and Pennebaker JW. Language style matching predicts relationship initiation and stability. *Psychological Science*. 2011, 22(1): 39–44.

13. Cooper JC, Dunne S, Furey M, and O'Doherty JP. Neural representations of reward in interpersonal attraction. Program No. 129.13.2010 Neuroscience Meeting Planner. San Diego, Calif.: Society for Neuroscience, 2010. Online.

Chapter 8: Making Love Last

1. Kleiman DG. Monogamy in mammals. *Quarterly Review of Biology*. 1977, 52:39–69.

2. Young LJ and Wang Z. The neurobiology of pair-bonding. *Nature Neuroscience*. 2004, 7:1048–54.

3.   Aragona BJ, Liu Y, Yu YJ, Curtis JT, Detwiler JM, Insel TR, and Wang Z. Nucleus accumbens dopamine differentially mediates the formation and maintenance of pair bonds. *Nature Neuroscience.* 2005, 9:133–39.

4.   Hinde K, Maninger N, Mendoza SP, Mason WA, Rowland DJ, Wang GB, Kukis D, Cherry SR, and Bales KL. D1 dopamine receptor binding potential as a function of a pair-bond status in monogamous titi monkeys (Callicebus cupreus). Program No. 903. 7/JJJ22. 2010 Neuroscience Meeting Planner. San Diego, Calif.: Society for Neuroscience, 2010. Online.

5.   Curtis JT, Liu Y, Aragona BJ, and Wang Z. Dopamine and monogamy. *Brain Research.* 2006, 1126(1): 76–90.

6.   Snowdon CT, Pieper BA, Boe CY, Cronin KA, Kurian AV, and Ziegler TE. Variation in oxytocin is related to variation in affiliative behavior in monogamous, pairbonded tamarins. *Hormones and Behavior.* 2010, 58(4): 614–18.

7.   Marazziti D and Canale D. Hormonal changes when falling in love. *Psychoneuroendocrinology.* 2004, 29(7): 931–36.

8.   Emanuele E, Politi P, Bianchi M, Minoretti P, Bertona M, and Geroldi D. Raised plasma nerve growth factor levels associated with early-stage romantic love. *Psychoneuroendocrinology.* 2006, 31(3): 288–94.

9.   Kim W, Kim S, Jeong J, Lee K-U, Ahn K-J, Chung Y-A, Hong K-Y, and Chae J-H. Temporal changes in functional magnetic resonance imaging activation of heterosexual couples for visual stimuli of loved ones. *Psychiatry Investigations.* 2009, 6(1): 19–25.

10.   Acevedo BP, Aron A, Fisher HE, and Brown LL. Neural correlates of long-term intense romantic love. *Social Cognitive and Affective Neuroscience.* 2011, 6(1). (pages?)

11.   Walum H, Westberg L, Henningsson S, Neiderhiser JM, Reiss D, Igl W, Ganiban JM, Spotts EL, Pedersen NL, Eriksson E, and Lichtenstein P. Genetic variation in the vasopressin receptor 1A gene (AVPR1A) associates with pair-bonding behavior in humans. *Proceedings of the National Academy of Sciences.* 2008, 105(37): 14153–56.

12.   Harkey SL, Brock AB, Kuehnmunch M, Krzywosinski T, Mitry MA, and Aragona BJ. Opioid regulation of pair bonding in the monogamous prairie vole. Program No. 387. 4/DDD6. 2010 Neuroscience Meeting Planner. San Diego, Calif.: Society for Neuroscience, 2010. Online.

Chapter 9: Mommy (and Daddy) Brain

1.   McEwen, B. Meeting report: Is there a neurobiology of love? *Molecular Psychiatry.* 1997, 2(1): 15–16.

2.   Ferris CF, Kulkarni P, Sullivan JM Jr., Harder JA, Messenger TL, and Febo M. Pup suckling is more rewarding than cocaine: Evidence from functional magnetic resonance imaging and three-dimensional computational analysis. *Journal of Neuroscience.* 2005, 25(1): 149–56.

3.   Swain JE, Lorberbaum JP, Kose S, and Strathearn L. Brain basis of early parent-infant interactions: Psychology, physiology, and *in vivo* functional neuroimaging studies. *Journal of Child Psychology and Psychiatry.* 2007, 48(3–4): 262–87.

4.   Champagne F, Diorio J, Sharma S, and Meaney MJ. Naturally occurring variations in maternal behavior in the rat are associated with differences in

estrogen-inducible central oxytocin receptors. *Proceedings of the National Academy of Sciences*. 2001, 98(22): 12736–41.

5. Champagne FA, Chretien P, Stevenson CW, Zhang T-Y, Gratton A, and Meaney MJ. Variations in nucleus accumbens dopamine associated with individual differences in maternal behavior in the rat. *Journal of Neuroscience*. 2004, 24(17): 4113–23.

6. Shahrokh DK, Zhang T-Y, Diorio J, Gratton A, and Meaney MJ. Oxytocin-dopamine interactions mediate variations in maternal behavior in the rat. *Endocrinology*. 2010, 151(5): 2276–86.

7. Ross HE and Young LJ. Oxytocin and the neural mechanisms regulating social cognition and affiliative behavior. *Frontiers in Neuroendocrinology*. 2009, 30(4): 534–47.

8. Feldman R, Weller A, Zagoory-Sharon O, and Levine A. Evidence for a neuroendocrinological foundation for human affiliation. *Psychological Science*. 2007, 18(11): 965–70.

9. Bartels A and Zeki S. The neural correlates of maternal and romantic love. *Neuroimage*. 2004, 21(3): 1155–66.

10. Leibenluft E, Gobbini MI, Harrison T, and Haxby JV. Mothers' neural activation in response to pictures of their children and other children. *Biological Psychiatry*. 2004, 56:225–32.

11. Swain JE, Leckman JF, Mayes LC, Feldman R, Constable RT, and Schultz RT. Neural substrates and psychology of human parent-infant attachment in the postpartum. *Biological Psychiatry*. 2004, 55:153S.

12. Noriuchi M, Kikuchi Y, and Senoo A. The functional neuroanatomy of maternal love: Mother's response to infant's attachment behaviors. *Biological Psychiatry*. 2008, 63(4): 415–23.

13. Kim P, Leckman JF, Mayes LC, Feldman R, Wang X, and Swain JE. The plasticity of human maternal brain: Longitudinal changes in brain anatomy during the early postpartum period. *Behavioral Neuroscience*. 2010, 124(5): 695–700.

14. Gordon I, Zagoory-Sharon O, Leckman JF, and Feldman R. Oxytocin and the development of parenting in humans. *Biological Psychiatry*. 2010, 68(4): 377–82.

15. Feldman R, Gordon I, Schneiderman I, Weisman O, and Zagoory-Sharon O. Natural variations in maternal and paternal care are associated with systematic changes in oxytocin following parent-infant contact. *Psychoneuroendocrinology*. 2010, 35(8): 1133–41.

Chapter 10: Might as Well Face It, You're Addicted to Love

1. Ke$ha, "Your Love Is My Drug." Spin Doctors, "I Can't Kick the Habit." Barry White, "Can't Get Enough of Your Love, Babe." Diana Ross, "Love Hangover." The Four Tops, "Baby, I Need Your Lovin'." Mariah Carey, "Can't Let Go."

2. Beydoun SR, Wang JT, Levine RL, and Farvid A. Emotional stress as a trigger of myasthenic crisis and concomitant Takotsubo cardiomyopathy: A case report. *Journal of Medical Case Reports*. 2010, 4:393.

3. Frascella J, Potenza MN, Brown LL, and Childress AR. Shared brain vulnerabilities open the way for nonsubstance addictions: Carving addiction at a new joint? *Annals of the New York Academy of Sciences*. 2010. 1187:294–315.

4. Ibid.

5. Schultz W. Multiple reward signals in the brain. *Nature Reviews Neuroscience*. 2000, 1:199–207.

6. Insel TR. Is social attachment an addictive disorder? *Physiology and Behavior*. 2003, 79:351–57.

7. Szalavitz M. The "mommy brain" is bigger: How love grows a new mother's brain. *Time*. 21 October 2010.

8. Kinsley CH and Meyer EA. The construction of the maternal brain: Theoretical comment on Kim et al. (2010). *Behavioral Neuroscience*. 2010, 124(5): 710–14.

9. Fisher HE, Brown LL, Aron A, Strong G, and Mashek D. Reward, addiction and emotion regulation systems associated with rejection in love. *Journal of Neurophysiology*. 2010, 104(1): 51–60.

10. Davis JF, Loos M, Di Sebastiano AR, Brown JL, Nehman MN, and Coolen LM. Lesions of the medial prefrontal cortex cause maladaptive sexual behavior in male rats. *Biological Psychiatry*. 2010, 67(12): 1199–204.

11. Edwards S and Self DW. Monogamy: Dopamine ties the knot. *Nature Neuroscience*. 2006, 9(1): 7–8.

12. Garcia JR, MacKillop J, Aller EL, Merriwether AM, Wilson D, and Lum JK. Associations between dopamine D4 receptor gene variation with both infidelity and sexual promiscuity. *PLoS ONE*. 2010, 5(11): e14162.

Chapter 11: Your Cheating Mind

1. Spring JA and Spring M. *After the Affair: Healing the Pain and Rebuilding Trust When a Partner Has Been Unfaithful*. 1997. Harper Paperbacks, New York.

2. Zuk M. *Sexual Selections: What We Can and Can't Learn about Sex from Animals*. 2003. University of California Press, Berkeley.

3. Harlow JM. *Recovery from the Passage of an Iron Bar through the Head*. 1869. David Clapp & Son, New York.

4. Pitkow LJ, Sharer CA, Ren X, Insel TR, Terwilliger EF, and Young LJ. Facilitation of affiliation and pair-bond formation by vasopressin receptor gene transfer into the ventral forebrain of a monogamous vole. *Journal of Neuroscience*. 2001, 21(18): 7392–96.

5. Walum H, Westberg L, Henningsson S, Neiderhiser JM, Reiss D, Igl W, Ganiban JM, Spotts EL, Pedersen NL, Eriksson E, and Lichtenstein P. Genetic variation in the vasopressin receptor 1A gene (AVPR1A) associates with pair-bonding behavior in humans. *Proceedings of the National Academy of Sciences*. 2008, 105(37): 14153–56.

6. Holden C. Why men cheat. *Science*. 2 September 2008. http://news.sciencemag.org/sciencenow/2008/09/02-01.html.

7. No author listed. Infidelity: It's all in the genes. *SkyNews*. 3 September 2008. http://news.sky.com/skynews/Home/Health/Unfaithfulness-infidelity-men-woman-genes-marriage/Article/200809115091928.

8. Garcia JR, MacKillop J, Aller EL, Merriwether AM, Wilson DS, and Lum JK. Associations between dopamine D4 receptor gene variation with both infidelity and sexual promiscuity. *PLoS One*. 2010, 5(11): e14162.

9. Ophir AG, Phelps SM, Sorin AB, and Wolff JO. Social but not genetic monogamy is associated with greater breeding success in prairie voles. *Animal Behaviour*. 2008, 75(3): 1143–54. (Highlighted in *Nature*. 2008, 451:617.)

10. Ophir AG, Gessel A, Zheng DJ, and Phelps SM. A socio-spatial memory neural circuit predicts male monogamy in the field. Program No. 387.10/EEE2. 2010 Neuroscience Meeting Planner. San Diego, Calif.: Society for Neuroscience, 2010. Online.

11. Weymouth WL, Richman E, and Phelps SM. Evolutionary remains: Differences in heritability of forebrain V1aR. Program No. 387. 9/EEE1. 2010 Neuroscience Meeting Planner. San Diego, Calif.: Society for Neuroscience, 2010. Online.

12. Cho MM, DeVries AC, Williams JR, and Carter SC. The effects of oxytocin and vasopressin on partner preferences in male and female prairie voles (*Microtus ochrogaster*). *Behavioral Neuroscience*. 1999, 113(5): 1071–79.

13. Bales KL and Carter CS. Developmental exposure to oxytocin facilitates partner preferences in male prairie voles (*Microtus ochrogaster*). *Behavioral Neuroscience*. 2003, 117(4): 854–59.

Chapter 12: My Adventures with the O-Team

1. Herbenick D, Reece M, Schick V, Sanders SA, Dodge B, and Fortenberry JD. An event-level analysis of the sexual characteristics and composition among adults ages 18 to 59: Results from a national probability sample in the United States. *Journal of Sexual Medicine*. 2010, 7 (Supplement 5): 346–61.

2. Georgiadis JR, Simone Reinders AA, Paans AM, Renken R, and Kortekaas R. Men versus women on sexual brain function: Prominent differences during tactile genital stimulation, but not during orgasm. Human Brain Mapping. 2009, 30(10): 3089–101.

3. Komisaruk BR and Whipple B. Functional MRI of the brain during orgasm in women. *Annual Review of Sex Research*. 2005, 16:62–86.

4. Arnow BA, Desmond JE, Banner LL, Glover GH, Solomon A, Polan ML, Lue TF, and Atlas SW. Brain activation and sexual arousal in healthy, heterosexual males. *Brain*. 2002, 125(Part 5): 1014–23.

5. Georgiadis JR, Reinders AA, Van der Graaf FH, Paans AM, and Kortekaas R. Brain activation during human male ejaculation revisited. *Neuroreport*. 2007, 18(6): 553–57.

6. Georgiadis et al. Men versus women on sexual brain function.

Chapter 13: A Question of Orientation

1. Derfner J. *Swish: My Quest to Become the Gayest Person Ever and What Ended Up Happening Instead*. 2009. Broadway Books, New York.

2. Wiltgen SM. A historical review of research related to the neurobiology of homosexuality. Program No. 21. 8/NNN6. 2010 Neuroscience Meeting Planner. San Diego, Calif.: Society for Neuroscience, 2010. Online.

3. LeVay S. *Gay, Straight and the Reason Why: The Science of Sexual Orientation*. 2010. Oxford University Press, New York.

4. Grosjean Y, Grillet M, Augustin H, Ferveur JF, and Featherstone DE. A glial amino-acid transporter controls synapse strength and courtship in Drosophila. *Nature Neuroscience*. 2008, 11(1): 54–61.

5. Park D, Choi D, Lee J, Lim DS, and Park C. Male-like sexual behavior of female mouse lacking fucosemutarotase. *BMC Genetics*. 2010, 7(11): 62.

6. Wallen K and Parsons WA. Sexual behavior in same-sexed nonhuman

primates: Is it relevant to understanding human sexuality? *Annual Review of Sex Research.* 1997, 8:195–223.

7. Roselli CE, Larkin K, Resko JA, Stellflug JN, and Stormshak F. The volume of sexually dimorphic nucleus in the ovine medial preoptic area/anterior hypothalamus varies with sexual partner preference. *Endocrinology.* 2004, 145(2): 478–83.

8. Mustanski BS, Dupree MG, Nievergelt CM, Bocklandt S, Schork NJ, and Hamer DH. A genomewide scan of male sexual orientation. *Human Genetics.* 2004, 116(4): 272–78.

9. Hu S-H, Wei N, Wang Q-D, Yan L-Q, Wei E-Q, Zhang M-M, Hu J-B, Huang M-L, Zhou W-H, and Xu Y. Patterns of brain activation during visually evoked sexual arousal differ between homosexual and heterosexual men. *American Journal of Neuroradiology.* 2008, 29:1890–96.

10. Bao A-M and Swaab DF. Sex differences in the brain, behavior and neuropsychiatric disorders. *Neuroscientist.* 2010, 16(5): 550–65.

11. Blanchard R. Quantitative and theoretical analyses of the relation between older brothers and homosexuality in men. *Journal of Theoretical Biology.* 2004, 230(2): 173–87.

12. Rahman Q. The neurodevelopment of human sexual orientation. *Neuroscience and Biobehavioral Reviews.* 2005, 29(7): 1057–66.

13. Savic I, Berglund H, and Lindström P. Brain response to putative pheromones in homosexual men. *Proceedings of the National Academy of Sciences.* 2005, 102(20): 7356–61.

14. Berglund H, Lindström P, and Savic I. Brain response to putative pheromones in lesbian women. *Proceedings of the National Academy of Sciences.* 2006, 103(21): 8269–74.

15. Berglund H, Lindström P, Dhejne-Helmy C, and Savic I. Male-to-female transsexuals show sex-atypical hypothalamus activation when smelling odorous steroids. *Cerebral Cortex.* 2008, 18(8): 1900–908.

16. Zeki S and Romaya JP. The brain reaction to viewing faces of opposite- and same-sex romantic partners. *PLoS ONE.* 2010, 5(12): e15802.

Chapter 14: Stupid Is as Stupid Loves

1. Karremans JC, Verwijmeren T, Pronk TM, and Reitsma M. Interacting with women can impair men's cognitive functioning. *Journal of Experimental Social Psychology.* 2009, 45(4): 1041–44.

2. Rubin C. Beautiful girls make men stupid. *Tonic.* 4 September 2009, http:// blog.tonic.com/the-effects-of-pretty-girls/.

3. Rubin C. Why beautiful women (literally) make men dumber. *Excelle.* 18 September 2009, http://excelle.monster.com/news/articles/4085-why-beautiful-women-literally-make-men-dumber.

4. Dabbs JM and Dobbs MG. *Heroes, Rogues and Lovers: Testosterone and Behavior.* 2001. McGraw-Hill, New York.

5. Rupp HA, James TW, Ketterson ED, Sengelaub DR, Janssen E, and Heiman JR. Neural activation in women in response to masculinized male faces: Mediation by hormones and psychosexual factors. *Evolution of Human Behavior.* 2009, 30(1): 1–10.

6. Rupp HA, James TW, Ketterson ED, Sengelaub DR, Janssen E, and Heiman

JR. Neural activation in the orbitofrontal cortex in response to male faces increases during the follicular phase. *Hormones and Behavior.* 2009, 56(1): 66–72.

7. No author listed. Explore your dark side to win her over. *Men's Health.* http://www.menshealth.co.uk/sex/more/explore-your-dark-side-to-win-her-over.

8. Grayson A. Why nice guys finish last. ABC News. 19 June 2008, http://abcnews.go.com/Health/story?id=5197531&page=2.

9. Rupp HA, James TW, Ketterson ED, Sengelaub DR, Janssen E, and Heiman JR. The role of the anterior cingulate cortex in women's sexual decision making. *Neuroscience Letters.* 2009, 449(1): 42–47.

Chapter 15: There's a Thin Line between Love and Hate

1. De Dreu CKW, Greer LL, Handgraaf MJJ, Shalvi S, Van Kleef GA, Baas M, Ten Velden FS, Van Dijk E, and Feith SWW. The neuropeptide oxytocin regulates parochial altruism in intergroup conflict among humans. *Science.* 2010, 238(5984): 1408–11.

2. Zeki S and Romaya JP. Neural correlates of hate. *PLoS ONE.* 2008, 3(10): 1–8.

Chapter 16: The Greatest Love of All

1. Ramachandran VS, Hirstein WS, Armel KC, Tecoma E, and Iragui V. The neural basis of religious experience. *Society for Neuroscience Abstracts.* 1997, 23:1316.

2. McKay R. Hallucinating God? The cognitive neuropsychiatry of religious belief and experience. *Evolution and Cognition.* 2004, 10(1): 114–25.

3. Ramachandran VS and Blakeslee S. *Phantoms in the Brain: Probing the Mysteries of the Human Mind.* 1998. Morrow, New York.

4. Cook CM and Persinger MA. Experimental induction of the "sensed presence" in normal subjects and an exceptional subject. *Perceptual and Motor Skills.* 1997, 85(2): 683–93.

5. Beauregard M and Paquette V. Neural correlates of a mystical experience in Carmelite nuns. *Neuroscience Letters.* 2006, 405(3): 186–90.

6. Beauregard M, Courtemanche J, Paquette V, and St-Pierre EL. The neural basis of unconditional love. *Psychiatry Research: Neuroimaging.* 2009, 172(2): 93–98.

Conclusion: A Brave New World of Love

1. Young LJ. Being human: Love: Neuroscience reveals all. *Nature.* 2009, 457(8): 148.

2. Tierney J. Anti-love drug may be ticket to bliss. *New York Times.* 12 January 2009.

3. Robbins T. *Still Life with Woodpecker.* 1990. Bantam, New York.

Index

—

Page numbers in *italics* refer to illustrations.

About the Author

—

Kayt Sukel earned a BS in cognitive psychology from Carnegie Mellon University and an MS in engineering psychology from the Georgia Institute of Technology. She is a passionate traveler and science writer, and her work has appeared in the *Atlantic Monthly,* the *New Scientist, USA Today, The Washington Post, Islands, Parenting, The Bark, American Baby,* and the *AARP Bulletin.* She is a partner at the award-winning family travel website Travel Savvy Mom (www.travelsavvymom.com) and is also a frequent contributor to the Dana Foundation's many science publications (www.dana.org). Much of her work can be found on her website, kaytsukel.typepad.com, including stories about out-of-body experiences, computer models of schizophrenia, and exotic travel with young children. She lives outside Houston and frequently overshares on Twitter as @kaytsukel.

02-13-12